OTHER BOOKS BY PETER WOHLLEBEN

The Hidden Life of Trees:
What They Feel, How They Communicate—Discoveries From a Secret World

The Inner Life of Animals:
Love, Grief, and Compassion—Surprising Observations of a Hidden World

The Secret Wisdom of Nature:
Trees, Animals, and the Extraordinary Balance of All Living Things—Stories From Science and Observation

The Hidden Life of Trees:
The Illustrated Edition

GREYSTONE BOOKS

PETER WOHLLEBEN

TRANSLATED BY **JANE BILLINGHURST**

THE HEARTBEAT OF TREES

EMBRACING OUR ANCIENT BOND WITH FORESTS AND NATURE

GREYSTONE BOOKS
Vancouver/Berkeley

First published in English by Greystone Books in 2021
Originally published in German as *Das geheime Band zwischen Mensch und Natur: Erstaunliche Erkenntnisse über die 7 Sinne des Menschen, den Herzschlag der Bäume und die Frage, ob Pflanzen ein Bewusstsein haben* by Peter Wohlleben

21 22 23 24 25 5 4 3 2 1

Greystone Books Ltd.
greystonebooks.com

Cataloguing data available from Library and Archives Canada
ISBN 978-1-77164-689-5 (cloth)
ISBN 978-1-77164-690-1 (epub)

Editing by Jane Billinghurst
Copyediting by Tracy Bordian
Proofreading by Alison Strobel
Indexing by Stephen Ullstrom
Jacket design by Nayeli Jimenez and Jessica Sullivan
Text design by Nayeli Jimenez
Jacket photograph by Abstract Aerial Art/Getty Images
Author photograph by Miriam Wohlleben
Endsheet photograph by Anna Om/Shutterstock

Printed and bound in Canada on ancient-forest-friendly paper by Friesens

Greystone Books gratefully acknowledges the Musqueam, Squamish, and Tsleil-Waututh peoples on whose land our office is located.

Greystone Books thanks the Canada Council for the Arts, the British Columbia Arts Council, the Province of British Columbia through the Book Publishing Tax Credit, and the Government of Canada for supporting our publishing activities.

Canada

CONTENTS

INTRODUCTION

IN RECENT YEARS, all over the world, there has been renewed interest in ways to immerse ourselves in nature. Forest bathing, for instance, has emerged as a therapeutic practice—in Japan, you can even get a prescription for it. At the same time, forests continue to be clear-cut with no thought given to the consequences. This reckless removal of trees fuels climate change. Faced with these contradictions, it can be hard for us to reclaim our place in the natural world. None of us deliberately sets out to destroy the environment, and yet we are all caught up in our consumer-oriented lives.

Assigning blame or giving in to despair, however, are far from helpful. Pointing an accusing finger at an apocalypse just waiting to happen, at a tipping point beyond which there appears to be no return to a stable climate, conjures up images of medieval inquisitions and couldn't be further from the positive encouragement we so desperately need right now.

And so, I invite you to join me in the forest instead, where we will discover that the ancient tie that binds humans and nature exists to this day and is as strong as ever.

Don't worry, our connection with nature is not so diminished that our only hope of long-term survival lies with modern technology. On this journey into the forest, you'll be amazed at how well your senses function. You are, for example, better at detecting some smells than dogs are. We'll discover that electrical activity in trees gives spiders goosebumps and explore a well-stocked natural pharmacy available not only to animals but also to you. And while you're exploring, you'll be surrounded by a cocktail of chemical communication that will strengthen your circulatory and immune systems.

Many people no longer notice these wonders. Not because our senses have atrophied—they all still work just fine, as the many different examples in this book will show you—but rather because of a strange philosophical and scientific worldview that erects unnecessary barriers between us and our fellow life-forms. Over here we have people, and over there we have nature. Over here reason runs the show, while over there a sophisticated, mindless, and apparently almost mechanical system runs its course.

The realization that we are still a part of this wonderful system and that we function according to the same rules as all other species is, thank goodness, gradually making headway. And it's only when it comes to the fore that conservation can be effective—that is to say, when we realize that what we are conserving is not just other forms of life but, first and foremost, ourselves.

1

WHY IS THE FOREST GREEN?

More and more people are taking delight in nature and not only want to see the forest but also to experience it as intensely as they can—and I am one of them. We often envy animals for the immediate, clear feedback their senses give them, but how aware are we of our own senses? After living for centuries in a human-oriented world that robs us of the daily necessity of keeping a wary eye on nature, what skills do we still possess?

If we are to believe the multitude of reports that compare the amazing skills of animals with our own, as a species we don't have much to offer other than brain power. In matters of the senses, we fall short when measured against almost all of our fellow creatures. Sometimes we even seem to relish our role as evolutionary losers. And so, the bond between people and nature appears to be ruptured beyond repair, and all there is left for us to do is to peer enviously at the amazing abilities animals possess.

We couldn't be more wrong. We are completely capable of engaging effectively with the world in which we live. It wasn't so long ago that our ancestors had to fight their way through forests, registering the presence of every possible danger or potential prey quickly enough to act. And because the blueprint for making humans has not changed since then, we can console ourselves with the thought that all our senses are still intact. The only thing missing is a bit of practice—and here we can catch up.

Let's first consider vision and ask a seemingly simple question: Why do we see trees in color?

We know we feel relaxed when we look at green trees. A shady green view even improves our health. But why do we see the color green in the first place? After all, this is not a skill most other mammals share with us. Their world is restricted to a narrow range of colors. Take the highly intelligent dolphin. Like many marine mammals, dolphins see the world in black and white, because their retinas contain just one type of cone (cones are cells that make it possible to see color). To distinguish between two colors, you need at least two different types of cones. Paradoxically, the one cone dolphins and other similar animals have is for the color green. This one cone allows them to distinguish between various levels of brightness, but that is all. Dolphins can't even process blue light, which not only colors the surface of the ocean but also reaches way down into the depths.

OUR FOUR-LEGGED DOMESTIC companions, such as dogs and cats, and wild forest animals, such as deer or wild pigs, do considerably better than dolphins in the color department. In these mammals, the green cones are joined by blue cones, and this combination allows them to see a limited range of colors—although the various shades of red, yellow, and green all run together and look the same to them. Having both green and blue cones is still not enough, however, to be able to see the color green. To do that, you also need to have cones that are sensitive to red light—as humans and many other primates do. And so, even though the color green calms our minds and promotes healing processes in us, it plays no role in the lives of most mammals.

But why do you need cones that are sensitive to green light and cones that are sensitive to red light in order to see the color green? This has to do with the wavelengths of light. Shades of blue have short wavelengths, and shades of green and red have longer wavelengths. If you are an animal that has only blue cones and green cones, whether the light entering your eye is green, yellow, or red, these "longer wave" shades stimulate only the green cones when they hit your retina, and all these colors look the same to you. Light with short wavelengths stimulates your blue cones, and light with longer wavelengths doesn't affect blue cones at all. That is why an animal that has cones for only blue and green can, strictly speaking, only distinguish between "blue" and "not blue."

It is only when another type of cone is added, one that is sensitive to another range of long-wave light, that a forest can be seen as green. And, wonder of wonders, we humans possess just such a cone in our retina.[1] It is sensitive to red light, and only when these three cones are functioning can we clearly distinguish whether the tree is green, yellow, or red. There's a reason the little LED lights in your computer or television screen are composed of minuscule blue, green, and red dots. If you have these colors, you can create any color you want.

Seeing forests as green, therefore, is a special skill if you are a mammal. But why, among mammals, have we humans developed this ability? Researchers suspect it has less to do with the color green and more to do with the color red. For example, many fruits found among the leaves of trees and bushes are red when ripe. We are not the only ones with our sights set on these. Many birds also have their eyes on them, and birds see red even better than we do. Plants have reacted to the situation: fruit that is eaten by mammals tends to be greenish-red when ripe, whereas fruit favored by birds is bright red.[2]

It makes sense, then, that we can see red, but why is it that we find green so beautiful? In fact, why do we notice it at all? You might think this is an odd question. We have the cones for green, and so it seems hardly surprising that we notice this color everywhere in the forest all the time. But that doesn't necessarily have to be the case. Consider the color blue. Our ancestors probably didn't notice blue at all or, if they did, they considered it unimportant. Lazarus Geiger, an nineteenth-century German linguist, discovered that in many ancient languages there is no word for blue. Homer, an ancient Greek writer about whom we know very little, probably lived about eight hundred years before the birth of Christ. He described the color of the ocean as "wine-dark," and texts from later centuries categorized blue as a shade of green. It was only with the development of and trade in blue fabric that the concept of "blue" was born. Since then, we have separated it out as a color in its own right and been consciously aware of it.

SO, DO WE see some colors only because there is a cultural reason to do so? Or, to put it another way, can we see blue only because we have a word to describe it? Jules Davidoff, professor of psychology at Goldsmiths, University of London, published the results of an impressive experiment on this subject. He and his team traveled to visit the Himba, a Namibian tribe that has no word for blue. On a computer monitor, he showed his test subjects twelve squares arranged in a circle. Eleven of the squares were green and the twelfth was very clearly blue. The Himba had great difficulty finding the blue square. Then he reversed the experiment. Davidoff showed people whose mother tongue was English another circle of twelve squares, this time all green. One of the squares had a tiny tinge of yellow in it that even I could not see. (You can take the test for yourself on the internet. The link is in the notes at the end of this book.[3]) English speakers had considerable difficulty finding the square in question. Not the Himba, however. They might not have a word for blue, but they have many more words for green than we do. This means they can describe even the smallest variations of color in green, and this is clearly what makes it easier for them to immediately identify the slightly differently colored square in the experiment.

Clues that the ability to see color is closely tied to culture also exist in countries where European languages are spoken. People whose mother tongue is Russian recognize different shades of blue far more quickly than non-Russian speakers, because Russian makes a clearer distinction between light blue and dark blue than other languages. A research team headed by New York psychologist Jonathan Winawer discovered that coworkers who spoke Russian were better at distinguishing shades of blue than their English-speaking colleagues.[4]

UNFORTUNATELY, I KNOW only of studies into the color blue. As a forester, I am, of course, interested in what's going on with the color green. When I look out my office window at the clearing around the forest lodge where my wife, Miriam, and I live, I see infinite variations on the color green. The blue-gray green of the lichens on the old birch tree; the yellowish green of the wintery grasses; the vibrant blue-green of the needles on the branches of the tall Douglas-firs; the warm, yellow-gray green film of algae growing on the bark of young beech trees—all of that is green to me. I certainly notice the differences between the various plants and their component parts, and there are descriptive terms in English such as pine green, shamrock green, and sage green, but these combinations are rarely used in everyday speech. Today, we tend to use less precise descriptors, such as light green or dark green.

A strong argument can be made that long ago our ancestors *were* able to distinguish between many different shades of green and red. If, as I explained earlier, recognizing red was important for our survival (because it meant we could find ripe fruit), then the same could be said for all the various gradations in color from green to yellow. How else could our forebears pick out ripe yellow corn cobs when the lush greens of plots they had so laboriously tended all year faded in the fall as the plants began to dry out and wither away? Or find fruits that changed from green (unripe) to yellow or red to indicate that they were ready to be picked?

A look back even further into the past shows how important it was to be able to make these distinctions. If an animal was wounded on the hunt, hunters could follow its trail only if they could clearly see drops of red blood on green grass. This ability to spot blood, incidentally, explains why one of the prerequisites when I applied for a job with the forest service—a job that, at the time, automatically included hunting animals—was full color vision.

TODAY, WE KNOW that red–green color blindness is genetically determined, just like the ability to see the color green. And yet, if the culture you are from affects how you see blue even though you have blue-sensitive cones in your eyes, it seems to me that the ability to see green is not something we should take for granted, either.

Writing is a good example of how much culture influences people's sensory perception. When you see the characters on this page, they form words with meaning, but Japanese characters would probably elicit a very different response—you might wonder how these symbols could ever give rise to mental images. Something similar happens with our sense of taste. Depending on the culture, the same food can be experienced as either disgusting or delicious, and you don't have to travel very far to see what I mean. In Sweden, for example, *surströmming,* fermented fish, is considered a delicacy. To me, however, it smells like fresh dog feces, and most tourists have an overwhelming urge to vomit as soon as the bulging can is opened.

Even if the ability to see green is determined by genetics rather than culture, that does not necessarily mean seeing green triggers a similar reaction in all of us. There is a lot of research that shows green, especially when we look at trees, affects our state of mind (I will look at this in more detail later). But might our reaction be determined by the historical era and culture in which we live? To answer this question, we would need more comparative studies, for instance, with people such as the Inuit, who rarely see green, or the Tuareg, who live in the Sahara, where the color you are most likely to encounter is some shade of brown. I am not currently aware of any such studies.

As fascinating as the subject of color is, the clarity with which we see things is much more important. And here, both genetics and the nature that surrounds us play a significant role. Sometimes, as I mentioned before, all we need to do is train our senses a bit to bring them up to speed.

DO YOU DISLIKE the idea of wearing glasses or perhaps just want to stop your eyesight from deteriorating? Then there is something you can do about it—at least when it comes to near-sightedness. I used to think that the tendency to near-sightedness was hereditary and that at some time in the future everyone on the planet would be wearing glasses. After all, these days almost no one's life still depends on whether they've spotted lions on the horizon in time to run away. In the absence of this kind of danger, it makes sense that the ability to see long distances deteriorates as evolutionary filters eliminate what is no longer necessary. Especially since we can overcome most limitations with appropriate aids.

Given this, are we all going to end up wearing glasses? Absolutely not, because science has recently discovered that near-sightedness is simply a case of our eyes adjusting well to seeing objects up close—think books and computers. The good news is that each of us can reverse, or at least put a stop to, this progression to near-sightedness. There is just one thing we need to do: go out into nature. As soon as our gaze drifts off into the distance, we are training our eyes to be far-sighted. If, however, we spend long hours at a desk, in low light and with our reading materials a short distance away, the advance of near-sightedness will continue.

These findings are the result of university studies that focused on children in East Asia. The change was particularly well documented in Taiwan during its rapid transformation into a more urban society. These days, 80 to 90 percent of Taiwanese high-school graduates need glasses, and 10 to 20 percent are struggling with visual impairments. What researchers first suspected might be genetic changes that would be passed down to the next generation were traced back to increased educational pressures and the accompanying loss of outdoor activities. Or, to put it another way, the benefits that came with modern society were turning young people into coach potatoes, and their sedentary lifestyles were the reason they needed to wear glasses.[5]

Near-sightedness caught up with me, too. When I was sixteen, my prescription was -2.5. That meant that the world more than 12 feet (3 meters) away from me was a complete blur. But my eyes did not stay that way. Unlike most of my fellow sufferers, my readings constantly improved and, after a few years, they hovered between -1 and a reading just above the level at which you no longer need to wear glasses. Even back then, I made the connection between the change in my eyesight and what I did for a living. For my work, I spent a lot of the day out in the forest evaluating trunks and crowns in stands of trees that were to be thinned, and I did all of this from a distance. I also spent a lot of my free time outside, repairing pasture fences or sawing logs for firewood.

Near-sightedness, therefore, is not an evolutionary adaptation as I feared, but simply our eyes adapting to seeing things up close, as they need to do for reading. Spending time out in nature and looking up or far away, at least when you are young, can improve or even prevent the problem.

THERE IS ANOTHER, completely different, way you can train your eyes. Maybe you've heard that dogs notice wild animals before we do? Contrary to what you might think, this often has nothing to do with scent, because the wind would have to be blowing directly at the dog. Rather, it's mostly because of movement, which our four-legged companions pick up with their peripheral vision. Our Münsterländer, Maxi, did this amazingly well from the window of a moving car.

Although I didn't know I was doing it, I've also trained myself in this skill over the course of my career. Wild animals are usually well camouflaged. The fur on deer is the same brown as the forest floor for a reason. But if a deer moves, I pick up on that out of the corner of my eye, even if the animal is some distance away. And I'm not alone in this, because, as it turns out, the human eye has an intriguing ability.

Our peripheral vision is actually very poor, and the resolution is so low that anything we see at the edge of our field of vision is blurry. And, as Dr. Laura Fademrecht and her research team at the Max Planck Institute for Biological Cybernetics in Tübingen, Germany, discovered, we can't even tell whether what we're looking at is a circle, a square, or one of the various other objects they used in their experiment. On its own, that discovery would not be particularly amazing were it not for the fact that when it comes to people, our peripheral vision can pick out many more details.

The researchers introduced life-sized stick action figures at the edge of the subjects' field of vision and had the figures make different movements. They made the figures wave, for example. Participants not only recognized these simplified shapes as people but were also able to judge immediately, based on their movements, whether they were being friendly or aggressive. From an evolutionary point of view, it is a distinct advantage to be able to immediately evaluate the intentions of approaching people. Peripheral vision, therefore, is hugely important to us as we interact with the outside world.[6]

You can try out this important skill even if you are in a city—supposedly as far from nature as you can possibly get. All those busy people moving about are great test subjects for your peripheral vision.

I GUESS IT'S not particularly surprising that our eyes still work extremely well, even if a closer investigation by scientists reveals a hidden skill. But what about our ears? Our sense of hearing is commonly thought of as weak compared with that of other members of the animal world, some might even say it has deteriorated from what it once was. But is that really true?

2

GIVING YOUR HEARING A WORKOUT IN NATURE

CAN YOU HEAR golden-crowned kinglets when they sing? These tiny birds weigh less than two-tenths of an ounce (just under 6 grams), and their song is pitched so high that it serves as an excellent test for your hearing. Their quiet *tsee-tsee-tsee* sounds almost like the few seconds of high-pitched buzzing a lot of us experience inside our ears from time to time. As we age, we lose our ability to hear higher-frequency sounds and the birds outside gradually fall silent.

Does this mean that our overall sense of hearing has atrophied? You might believe this if you compare our abilities with those of animals. Some internet sites go so far as to say that dogs can hear higher frequencies 100 million times better than we can.[7] That's a huge exaggeration, of course. It makes our ears sound completely useless. Are they really?

Let's look at the facts: People can hear sound waves with a frequency of 20 to 20,000 vibrations per second (20,000 Hz), whereas dogs hear from 15 to 50,000 vibrations per second (50,000 Hz). Our hearing is not that much worse. It's just that we can't hear anything above 20,000 Hz, a threshold above which the world is still filled with noise for dogs.

If we're going to make a comparison, a more meaningful one would be volume. Dogs pick up on quieter sound than we do simply because they have larger ear muscles and can point their ears toward the sound. It's easy to see how this works. Just cup your hands and place them behind your ears pointing forward. You'll find it makes a big difference. You can try this when you're out walking in the woods. Even if you're a long way away, you'll be able to hear the quieter birds or a deer slipping through the undergrowth.

However, one part of the claim that dogs and some other mammals hear better than we do because they can point their ears toward the sound and we can't turns out to be a myth. As far as the ear muscles go, it's absolutely correct. Only an estimated 10 to 20 percent of people can move their ears.[8] And the people who can, can't do much more than wiggle them and certainly can't point them forward. Recent research, however, shows that we have been too focused on external features. In fact, you and I can change the direction of our ears if we need to, but we make the adjustment internally. To do this, we need our eyes, as Kurtis Gruters, a graduate student in the department of neuroscience at Duke University in North Carolina, discovered.

Gruters tested sixteen subjects sitting in a completely darkened room. This allowed them to concentrate on colored LED lights that they were to track visually. Amazingly, the first thing that moved was not the subjects' eyes but their eardrums, which oriented toward the points of light. It took just 10 milliseconds for the subjects' eyes to follow.[9] You could, therefore, say that the eyes and the ears were directed to an object at about the same time. What's important here is not the time lag but the fact that we line up our auditory apparatus at all, an alignment that had never been noticed before. Even more surprising is that the test subjects' ears were oriented not to a sound but to an object they wanted to observe at with their eyes. Gruters' studies clearly show that we still have a thing or two to learn when it comes to our physical capabilities and, above all, that even our supposedly feeble and fixed ears can surprise us at any time with what they can do.

YOU CAN TRAIN your ears the same way you train your eyes—two senses that, as we've just seen, are inextricably entwined. All you need to do is keep your ears open and eavesdrop on nature. For example, I enjoy hearing the call of the black woodpecker. Perhaps that's because I know it relies on ancient beeches with wide trunks for its nesting cavities and it has become rarer for lack of suitable trees. Or maybe it's because of the bird's impressive size and its pretty bright-red feather cap. Whatever the reason, to this day it makes me happy every time I hear its cheery *croo-croo-croo*. I feel the same thrill when I hear the *crock-crock-crock* of the raven, which up until the end of the twentieth century was thought to have been extirpated from the Eifel, a low mountain range in western Germany where Miriam and I live, or when I hear the unmistakable call of the cranes that once again fly by the thousands over our forest lodge on their spring and fall migrations.

Because the calls of the cranes are among my most favorite sounds, I hear them even when, for most other people, they get lost in ambient noise. I register crane calls, for example, despite triple-glazing, insulated walls, and the nightly hum of the television. I jump up from the couch and run to the front door so I can enjoy the sound at full volume outside.

It shouldn't be a problem for anyone to hear nature more clearly. Think of other everyday sounds that, over time, you've learned to pick out. The ring of a cell phone, for example, or the sound when a message comes through. I'm always amused when I see my fellow travelers in trains or waiting in the station give an involuntary twitch when they hear those sounds from somewhere in their vicinity even when the volume is turned right down. As most people (including me) haven't customized their ring tones, cell phones of the same make all sound similar. Maybe that important call is for you?

If you tune your subconscious to the sounds of nature instead, it's easy to keep acoustic tabs on many of your fellow creatures.

3

YOUR GUT'S AMAZING SENSE OF SMELL

THE HUMAN NOSE seems to play almost no role when we're out in nature. At least, that's the impression I get on some of the walks I lead in the forest. When I ask people what it smells like under beeches or oaks, the first thing they do is inhale deeply. Up until that point, most of them have been exploring the forest with their eyes alone. They can describe the scents of the forest only after they have consciously taken in a "noseful" of their surroundings.

Just as we do with our sense of hearing, we dismiss our sense of smell as being really bad in comparison with animals, especially dogs. We laud our canine friends for their amazing skills in the sniffing department. A dog's sense of smell is generally thought to be a million times better than ours.[10] Moreover, 10 percent of a dog's brain is said to be devoted to processing scents, whereas only 1 percent of the human brain is set aside for this task.[11] I should make a brief observation here. As our brain is ten times the size of a dog's brain, the calculation of percentages is misleading because, in the overall scheme of things, we have the same brain area dedicated to decoding smells.

People make statements like these about dogs' abilities all the time, so it's hardly surprising that most of us don't think much about what our noses can do. The truth, however, as it always does, lies somewhere in between. It's true that dogs can sniff out many things much better than we can. The important question is, what kinds of things? And this was exactly the question Matthias Laska, professor of zoology at Linköping University, Sweden, asked. He tested fifteen different smells at the lowest levels dogs are aware of them. Then he tested human subjects at these same levels and, lo and behold, for at least five of these smells, humans outperformed his four-legged research subjects. When you think about this, the result is not totally unexpected because the five smells humans excelled at noticing came from the plant world; for example, from fruit.[12] Dogs are not that interested in fruit. What they want to sniff are things that are important to them. And those things are other dogs, deer, or squirrels and not apples, bananas, or mangoes.

I DON'T WANT to leave you with the wrong impression here. Overall, dogs certainly have a better sense of smell than we do—in their world smelling things is definitely more important than it is in ours. Our upright gait immediately puts us at a disadvantage. It's not really practical for us to put our noses to the ground to follow a scent trail. But then again, we don't need to. Our sense of smell is not here to help us track prey but to help us find delicious fruit hanging from tree branches or to find a mate.

When the scent from the opposite sex wafts over our 30 million olfactory cells, every once in a while, something "clicks." For example, a woman might become aware of a man with an unusually high level of testosterone. Or clear differences in DNA might attract us to a potential partner. Amazing as it sounds, a well-crafted perfume can have similar effects. We can mask our own body odors and make ourselves smell more attractive to others in ways they register on both a conscious and subconscious level.[13]

Using your nose to choose a partner is also common in the world of other mammals, and even here custom-designed perfumes come into play. It's a well-known fact that goats rely on them. Our billy goat, Vito, for example, changed how he smelled when mating time came around by spraying himself with his own proprietary perfume: his urine. He spritzed daily onto his front legs and into his mouth until we could smell him from a hundred yards away. The female goats found this most attractive. We, however, were less taken with the odor.

THE NOSE, HOWEVER, is not the only organ we smell with. We also have olfactory receptors in our bronchial tubes, which expand when they detect certain scents. And even our small intestine gets involved in sniffing our food. Researchers at Ludwig Maximilian University of Munich have discovered that the mucous membranes of our gut contain olfactory receptors for thymol and eugenol, the substances that give thyme and cloves their distinctive scents. The receptors for these compounds were always thought to be only in the nose. When the gut detects them, it releases chemicals and changes the way it moves.[14] The discovery is important because, in the natural world, we are exposed to a limited number of smells. The current flood of artificial compounds in perfumes, scented candles, and household cleaners can therefore cause intestinal discomfort and adversely affect our well-being.

WHEN SOME PEOPLE smell little or nothing out in the forest, they are not necessarily being inattentive. It could be that they have lost some or all of their ability to smell. And this is not unusual, as Dr. Sven Becker, a visiting researcher at the Ear, Nose, and Throat Clinic at the University of Munich, explained recently during a local radio program I was listening to. He estimated that 20 percent of the population in Germany has a compromised sense of smell, and 3 to 4 percent has lost this sense completely.[15]

And even when our noses are working exactly as they should, they will never be as important to us in experiencing the world as our eyes or ears because these latter two sense organs are much more important for how we communicate with one another. However, we should not underestimate our noses as organs of perception. We don't use them nearly enough when we set out to experience nature—but that is something we can change. The next time you step into a forest, why not stand still, close your eyes, and breathe in deeply. What does the forest smell like? Tangy? Sweet? Where are all those wonderful smells coming from anyway? It's time to go out and explore!

4

NATURE DOESN'T ALWAYS TASTE GOOD

I WAS ON A television talk show recently, and I had brought along something for the other guests to taste: branch tips from spruce and Douglas-fir. The spruce, the most common tree in Germany these days, is relatively well known. The Douglas-fir, less so. It is a conifer native to the west coast of North America, where these trees grow into impressive, ancient giants. In recent decades, more and more of them have been planted in Germany, but that was not the focus of the program. I had chosen the Douglas-fir branch tips because, in my opinion at least, they have the pleasantly tangy taste of candied orange peel. My fellow guests, the actor Axel Prahl and the cabaret performer Ilka Bessin, confidently bit into the soft, green tips—and quickly spat them out. They did not like the taste at all. A reaction they share with most people.

The flavors of the forest are mostly variations on tart and bitter and all the nuances in between. Those things we find delicious—that is to say, ripe nuts and berries—are usually in short supply and available for only a few weeks of the year. In spring, fresh growth and new leaves are tart, turning bitingly bitter and tough later in the season. A tree's cambium, a crystal-clear layer just under the bark that you can peel off with a pocketknife, is highly nutritious. It contains sugar and other carbohydrates, and tastes a bit like carrots, but on the whole, I would say it, too, tastes bitter. And that is the case for just about all the food you can find in the forest.

I'M FAIRLY CERTAIN that in the dim and distant past, the majority of the meals our ancestors ate tasted completely different from the meals we eat today. Our food and drink, just like the environment in which we live, have undergone evolutionary change. The only products that survive on supermarket shelves are the products people buy. And so, manufacturers develop foods that appeal most to our sense of taste. Their methods get ever-more sophisticated and match our desires ever-more precisely, which is also one of the reasons we find it so difficult not to reach for those particular foods. Sugar, salt, and fat—enrich that combination with other flavor enhancers, and we end up eating more than our body needs. In the process, we increasingly forget what natural, unprocessed food tastes like. I don't mean fruits and vegetables, because even these are being similarly transformed through selective breeding: always sweeter, always with fewer bitter compounds. In comparison with the rich variety of flavors available in nature, the food we eat all tastes more or less the same. Only certain particularly bitter or sour foods stand out, such as coffee or pickles.

Luckily, you can never completely spoil your sense of taste or dull your papillae, the taste centers on your tongue. A single papilla contains one hundred taste buds, and each taste bud contains one hundred sensory cells. These cells are not particularly long lived; they are renewed every ten days.[16] This means if you damage one of them when you are eating—for example, if you drink something that is too hot—your tongue heals relatively quickly.

If there are about one hundred papillae, that means we have about 10,000 taste buds. If that seems like way too many to you, just take a look at a horse's tongue, where you will find about 35,000 taste buds.[17] Why do horses need this many? There are hundreds of different kinds of grasses and weeds in every meadow, and quite a few of them are poisonous. A horse also has difficulty seeing anything that is directly in front of its lips because its huge, elongated head gets in the way. And if you can't see what you're eating, you have to rely on your tongue. If you're going to do this, the plant in question has to get into your mouth and then be removed quickly if it turns out it's not good to eat. Horses have mastered this perfectly. Miriam and I own two mares, and we never tire of watching them eat. If a bit of greenery is not to a horse's taste, it carefully moves it to the side of its mouth as it chews and from there the offending slip of green is eventually released back into the outside world by way of the horse's lips.

WHILE WE ARE on the subject of the tongue: it's not the only body part we taste with. First, let's return to our nose.

Currently, we know of about eight thousand substances in food that evaporate easily so we can smell them. Amazingly, we do most of this smelling as we exhale.[18] Three-quarters of our sensation of taste is based on what our nose picks up. You might know this from when you have a cold: food immediately tastes bland, and you no longer enjoy what you are eating. And so it makes perfect sense when you are out on your next walk in the woods to figure out the differences between tree species not by investigating the shape of their needles and leaves, but, like Ilka Bessin and Axel Prahl, by biting into a spruce tip to see which taste and scent compounds are to be found in the needles.

As I mentioned earlier, our search for taste sensors doesn't end in the mouth. We need to travel farther down the alimentary canal, into the gut. Just as the gut joins the nose in smelling things, it also joins the tongue in tasting things. It, too, contains sensors that were once thought to belong only in the nose. These cells are less easily tricked by sugary things than the cells in our palates. Normally, sugar, which the small intestine recognizes, triggers a release of hormones. Our consciousness registers this as a signal we are full. When we eat artificial sweeteners, however, this signal malfunctions and becomes weak and intermittent, which means our body craves more food. For this reason alone, "lite" products manufactured with sugar substitutes are not particularly effective if you want to lose weight.[19]

Thanks to modern cosmetics—and perfumed laundry products, fragrant candles, and the like—not only our noses and our palates, but also our guts are swamped with scents.[20] Wait a moment. Who eats cosmetics, laundry products, and candles? The answer is simple: we don't need to eat them for them to get inside us, because we absorb them through our skin and airways not only into our guts but also into every other corner of our body. And when we ingest flavor-enhanced foods, a veritable armada of artificial compounds drifts down our alimentary canals to overwhelm our sensory receptors.

According to the German Federal Institute for Risk Assessment, 2,700 (overwhelmingly artificially synthesized) aromas are added to food products.[21] That doesn't seem like very many when you compare them with the smells that are present in nature: to date we have identified about ten thousand of those. But numbers alone are deceiving. In our daily lives, we would encounter very few of these natural smells. After all, we don't taste every fruit in the world, only the ones that grow close to where we live—or at least that used to be the case before the advent of global trade.

Nowadays, our gut is overwhelmed by an unbelievable number of alien aromas. That can cause it to act up occasionally and lead to all sorts of intestinal complaints, because sensing aromas, depending on which ones they are, can, as I mentioned earlier, trigger the release of chemicals and change the way our gut moves.

WHAT HAS ALL this got to do with the forest? Well, we've adapted to the smells and tastes of this natural ecosystem, and we're just fine with them. Artificial additives, however, stress our systems unnecessarily. Therefore, it's a good idea to give your nose, palate, and gut a break every now and again by going out into the forest and spending a good long time there. Everything that flows over your senses when you are in the forest is exactly the kind of compound your body is made for. If you take along a snack of natural, minimally processed food without additives, then your time spent forest bathing will be even more beneficial.

5

TOUCH HELPS US THINK

WE'VE PAID TRIBUTE to four of what we typically think of as our five senses. The fifth, which we'll investigate in this chapter, is actually the most important: our sense of touch. When we think of touch, the first thing that comes to mind for most people is our fingers. There's a neat forest game you can play to experience its importance for yourself. The game works equally well for adults and children. Blindfold one person, then have someone else lead the blindfolded person through the trees. If you're the one being led, you'll notice right away how much trust you have to put in your partner, as with every step you anticipate a painful collision between your head and a gnarly tree trunk. The goal of the short walk is a randomly chosen tree that the person wearing the blindfold has to touch. The blindfolded person now has to run their fingers all over it. The mossy cushions on the spreading roots, the texture of the bark, tiny twigs, the diameter of the trunk—all of these come into play. Finally, the pair return. Spin the person wearing the blindfold around to disorient them, then remove the blindfold. The exciting part comes next. Will the person be able to find the tree now that he or she can see? Usually, it works astoundingly well and one thing quickly becomes clear: our hands translate what they feel into pictures.

IN 2014, A team of international scientists set up an experiment to see if there is a direct connection between touch and pictures—that is to say, our eyes. This is what they discovered: whenever the test subjects touched something with their fingers, their eyes stopped moving for a fraction of a second.[22] We are not aware of these minuscule pauses, but clearly they are long enough to allow the brain to focus better and process what has been touched.

AS AN ORGANISM, we have a vast number of sensory cells devoted to touch. Up to 600 million receptors are hidden in our skin, but there are also receptors in our muscles, tendons, and joints.[23] Together, these two different kinds of receptors, tactile (in the skin) and kinesthetic (within the body), are known as haptic receptors. It turns out they are necessary not only to establish the outer limits of our own bodies, but also to help us with mental concentration.

Martin Grunwald, leader of the Haptic Research Laboratory at the University of Leipzig, feels psychologists do not pay nearly enough attention to our sense of touch.[24] With this in mind, he researched the way people spontaneously touch their faces. We all do it. You might be doing it right now while reading this book. These movements are not for communication and, in most cases, we are not even aware of them. But that does not mean they serve no purpose, as Grunwald discovered. He measured the brain activity of test subjects while they tried to remember a sequence of haptic stimuli for five minutes. When he disturbed them with unpleasant noises, the subjects dramatically increased the rate at which they touched their faces. When the noises upset the rhythm of their brains and threatened to disrupt the subjects' concentration, self-touch helped them get their brain waves back on track. To put it another way: self-soothing grounded their minds.[25]

SEE, TOUCH, LEARN—THAT is a triad that can easily get out of sync in our modern world. The more information we get from our smartphones and televisions, the less we use our sense of touch. It's too early to say what the long-term consequences might be, but it certainly can't hurt to take remedial action right now. And this time I don't mean by taking a walk in the woods, at least not the normal kind of walk.

The next time you are walking outside, try touching a variety of objects. The feather lying by the side of the path is just waiting to be picked up. Even the slippery rock covered with algae offers unusual experiences of contact and motion. If you're worried about touching your clothes with dirty hands, use a small cushion of moss to wipe them off—it cleans really well, especially when it's damp. And as a bonus you can chalk up yet another tactile experience.

6

TRAINING YOUR SIXTH SENSE

In addition to the well-known five senses—vision, hearing, smell, taste, and touch—scientists have identified other ways of perceiving the world. Some animals, for example, are able to sense electrical fields or volcanic eruptions before they happen. In the catastrophic tsunami that hit Southeast Asia in 2004, this was observed multiple times. For example, when panicked water buffalo raced inland, residents of the area saw the buffalo leaving and so also sought safety on higher ground, escaping the deadly wave.[26]

What I find much more exciting is that people have this sense, too. It's exciting because it deepens our connection with nature, even when this is sometimes very painful. One example is sensitivity to the weather. When a deep low displaces an area of high pressure, I sometimes experience headaches and my gums hurt. That's extremely unpleasant, but luckily the pain passes within a couple of hours. I share this sensitivity to changes in the weather with about

50 percent of the population. Are you, perhaps, one of these people? It doesn't help that many scientists remain skeptical, but that doesn't make the symptoms any less real.

There is one basic way the weather affects us that scientists don't dispute, but it's also pretty obvious. When the weather gets colder, our body has to produce more warmth to maintain a temperature of almost 98.6°F (37°C), and when temperatures rise, we sweat to cool down. These processes are accompanied by a rise and fall in blood pressure, as well as a contraction and then expansion of blood vessels, and these changes alone can lead to tingling sensations in organs and limbs.

But it seems to me that this explanation doesn't go far enough. I'm sensitive to changes in the weather even when I'm inside all day, where the temperature stays the same and my body doesn't have to make any adjustments. The only thing that changes on those days is the air pressure. If the pressure falls outside, then it falls by exactly the same amount inside—after all my house is not airtight. The experiences of millions of people have no scientific explanation. At least not yet.

THE QUESTION IS whether this definition of the sixth sense is even close to adequate or whether our body has a lot more to offer. Something like a body sense—that would be sense number seven. Have you ever heard of a seventh sense? It wouldn't be surprising if you hadn't, because it's not often discussed, even though it's one of the most important senses you possess. It is not associated with any particular organ, yet you feel it. Right now. It's telling you where your body ends, whether you're sitting in a well-balanced position, whether the sofa you are sitting on is soft, and whether the book you're holding is heavy. In short, it is an interplay of many organs and nerve cells all the way up to the brain, which analyses and interprets all the data it receives.

But you don't have to be an organism with a large central nervous system to have a sense like this. Even plants have one. Even they, after all, sense gravity, and trees maintain the equilibrium of trunks that weigh many tons. For example, as soon as a beech notices its crown is getting out of alignment, it grows specialized reinforcing wood to shore up one side of the trunk so it doesn't bend any farther on that side. At the same time, it grows a different kind of wood on the opposite side that acts like the guylines on a tent to stop the trunk from leaning over any farther.

Our sense of balance is an integral part of the sense that stops us from falling over even when our eyes are closed. If this sense is compromised, perhaps as a result of a disease that has attacked the nervous system, you can't keep your balance even if you can see perfectly well.

BUT BACK TO the main topic of this chapter. Traditionally, when we talk about a sixth sense, we mean something that appears to transcend the senses or, at least, cannot be explained fully using today's scientific methods. The weather sensitivity I've just described is one example, but there are others, like the feeling something dangerous is about to happen. Although this ability is often dismissed as being an example of esoteric thinking—that is, something that cannot be proven and belongs to the intangible world of religious and spiritual experience—scientists from Washington University in St. Louis decided to find out if there might be something to it. We're talking about the sudden vague feeling people sometimes have that something is not right. The body sounds the alarm and, in the best-case scenario, the person becomes consciously aware of the threat and manages to escape. The sensations can also be less dramatic, such as the feeling of being watched—you turn around and see that someone is indeed looking at you. The key question here is where this vague feeling comes from.

The researchers devised an experiment to find out more about this mysterious sense. The test subject sat in front of a screen on which either a blue dash or a white dash appeared. After a while, the dash turned into an arrow, which, depending on the direction it was pointing, required the test subject to push one of two buttons. Sometimes, a split second before the subject was to push the button, a second, larger arrow appeared on the screen, pointing in the opposite direction. By that time, it was too late for the subject to change her mind about which button she was going to push. Does that sound as though it would be confusing? It's supposed to. The test subjects were unable to detect a pattern; however, the initial color of the dash did in fact indicate if such a last-minute change was likely to happen. Subconsciously, the subjects picked up on this. After a series of tests, some subjects could subconsciously predict whether the second arrow, the one that changed the direction, was going to appear or not.

While all this was going on, the subjects' brain waves were being measured and, interestingly, there was one region that was particularly active: the anterior cingulate cortex or ACC for short. There is a lot we still do not know about this portion of our brain, but it is clear, thanks in part to the results of this experiment, that it is a place where subtle clues from the environment reach our subconscious and are then translated into a conscious awareness of our surroundings and linked to emotions.[27]

Our sixth sense, therefore, is located behind our forehead, where it diligently processes all kinds of information about our environment. While you are reading these words, your brain is registering the temperature of the room you are in and the sounds and smells around you. None of this penetrates your conscious thoughts while you are reading, because that would be distracting, but the moment your ACC deduces from the mix of sensory stimuli that some kind of action needs to be taken, it draws your attention away from your reading by making you feel uneasy. At that point, your sixth sense reaches your conscious awareness and you often cannot explain why you react the way you do—after all, you were not aware of the signals coming from your environment and you often don't notice them after the fact.

YOUR SIXTH SENSE should be considered as something that physically exists and, when it comes to your relationship with the natural world, it still functions just fine. The only important thing is to give it some practice out there. The ACC can't work miracles. It can assess only what it knows, and it needs a great deal of experience and exposure to learn which forest sounds, which wind strengths, and which types of ground pose potential danger.

And on the subject of exposure, you only get that—you guessed it—if you go out into the forest as much as possible. And yet many people are afraid to go out into the woods alone. Isn't it quite dangerous to be out walking all by yourself?

7

FEAR IN THE FOREST

HAVE YOU SEEN the film *Jaws*? I have seen it more than once, and I've regretted that for years now, because it ruined swimming in the ocean for me. I know that sharks are rarely dangerous—a study done in California showed that 1 in 738 million people who go to the beach and swim in the ocean are attacked.[28] And even though there are fewer and fewer of these endangered creatures in the ocean, fear always swims alongside me. When there are a lot of other people in the water, it's sort of okay, but if I'm alone, I don't dare venture any deeper than a few feet. My intellect can run through as many facts as it wants, but my emotions refuse to listen.

The key here is to familiarize yourself with local conditions and know what you are dealing with. Most wild animals, even fearsome predators, prefer to give people a wide berth and do not attack unless threatened. They do not want to risk injuries, which could end up being life-threatening for them, because that could affect their ability to hunt. They also simply don't recognize us as food. And if they recognize us as enemies, their instinct is to flee rather than to approach us. The more time you spend out in the forests, the more you will get a feel for the animals that live there, how to act appropriately around them, and when you should—and should not—be concerned.

There are striking parallels between fears and allergies. Allergies arise because we have removed most of the threats that normally attack our immune systems. Medications such as antibiotics and, even more importantly, extreme hygiene mean that in many cases our bodies don't have to deal with tiny attackers such as viruses, bacteria, or worms, and don't have to break down the proteins they contain. But our immune systems must remain constantly on the alert—just in case. And because they mostly have nothing to do, they begin to turn their attention to less dangerous alien entities. This is the reason the pollen of grasses and trees brings on such severe allergic reactions—after all, they are mostly made up of protein.

In Germany, the concentration of pollen from birch trees, a species that is particularly likely to trigger an allergic reaction, is increasing year by year. There are a couple of reasons for this: not only do cities continue to plant birches despite warnings from medical professionals, but these trees also spread easily on their own. The birch is a pioneer species, which means it is one of the first trees to grow on surplus land. And there is no shortage of this: railroad embankments, the margins of industrial areas, green islands where highways intersect, or the roofs of abandoned buildings where birches eke out an existence despite the lack of soil. The wind is their ally, distributing their dust-fine pollen for many miles.

Ragweed is a good example of how far wind-borne pollen can travel. As many as 23 million Americans suffer allergic reactions to ragweed pollen, making it one of the most common plant allergies in the country. Unfortunately, the plant grows in every state but Alaska, and its pollen has been found up to 400 miles (650 kilometers) offshore and up to 2 miles (3.25 kilometers) in the air.

Tree pollen can travel even farther. One study in the southern United States found that pollen from loblolly pines can travel up to 1,800 miles (2,900 kilometers).[29] In years when a particularly large number of trees from any species bloom, there is so much pollen coming from forests that the landscape looks as though it is shrouded in fog. The long airborne journeys their pollen makes help trees avoid inbreeding. Pollen in spring air, therefore, is something completely normal. Pollen allergies, however, are something new. Are our bodies, now that they don't have to defend themselves against other dangers, gradually turning against the place we once called home?

And what about our state of mind? Something similar, it appears, is happening in our heads.

FROM PREHISTORY TO the nineteenth century, a walk in the woods really could be a dangerous undertaking. Not so much because of predatory animals, but more because of our fellow human beings. According to historical reports, as late as the 1870s, robber bands lurked along forest roads in the Eifel, where I live, waiting to attack wagons filled with food being sent from the rich city of Cologne to the starving people in the country. Of course, there were also wolves. They were mainly a threat to livestock and were therefore seen as a deadly danger, for who could survive without milk cows and oxen to pull plows? Even back then, however, there were not many reports of wolves attacking people. Nevertheless, fairy tales fixed the idea of the big bad wolf in the public imagination, and Europeans took their fears with them when they immigrated to North America.

And today? Forests have become—for the most part—extremely safe places to be. There are no longer any robber bands, and animal attacks (with the exception of domestic dogs and the odd cow or two when you have to cross a pasture) are almost unheard of. Most poisonous snakes and insects or large predators are either in short supply or prefer to avoid people if they can. And yet, many people are afraid when they are out alone in the woods.

Try it for yourself. If you don't feel a frisson of fear during the day, test yourself with a night hike. Our instincts kick in particularly strongly in the dark and shoot down any attempt reason makes to reassure us that everything is just fine. I admit, even I occasionally feel a touch of fear lurking in the back of my mind (just a touch). Luckily, I have spent enough time in the forest over the years that it never succeeds in gaining the upper hand.

So, I suggest you embark on the same kind of desensitization program you might undertake with allergies: nighttime hikes, carefully measured out in small doses, will break down fears while at the same time giving all those senses that are underused during the day a welcome workout.

8

WE ARE BETTER THAN WE THINK

IT WAS IMPORTANT to me, in the chapters so far, to point out that our perceptual skills are as strong as they ever were. Our senses are not inferior to those of animals, they are simply—as with all species—perfectly adapted to our needs. And this makes humans absolutely normal. Why do we constantly give ourselves short shrift when it comes to our senses? Why do we always compare ourselves only with species that can do something better than we can, instead of with those that we outperform in some way?

There's no question that certain species have specialized ways of perceiving the world. Take birds of prey, which can pick out detail up to four times better than we can. That means, for example, that they can spy mice on the ground even when they are flying miles up in the air. Some species, such as vultures or falcons, even have a kind of built-in telescope that enlarges a portion of their visual field so they have even more acute long-range vision.[30] Sharks have an unbelievably good sense of smell. They can detect fish blood in a concentration of just 1 part per 10 billion. I should mention here that they can't do this for human blood, despite all the alarmist reports to the contrary. We are not on their radar as potential prey and, therefore, they usually show absolutely no interest in us.

At the end of the day, every species has its own distinct abilities. Each has exactly the skills it needs to survive in its ecological niche. To come back to the first example I mentioned in this context, dogs need a good nose, like the one their wolf ancestors gave them, so they can track down prey. Vision and taste don't need to be quite as sophisticated in their world as they are in ours, but they have greater need of a sense of hearing than we do. They are perfectly adapted to their ecological niche, just as we are to ours. Therefore, comparisons make no sense. If you look at the world from the perspective of adaptation, there is no such thing as better or worse.

Our senses function just as they did for our ancestors thousands of years ago, and they allow us to focus our attention on our world, a world that does not consist primarily of desks, couches, and fast-food restaurants but of forests and open plains—at least that is the way it should be even today. We are perfectly equipped for those last two environments and could, at any time (after a couple of weeks of training), hold our own in the wild.

I BELIEVE THAT many nature lovers harbor a deep desire not to be the ones in charge on this planet. With daily news about environmental destruction and apocalyptic messages about climate change, we come across as so brutal in our dealings with every one of our fellow creatures that it seems the connections we have with all the other species that inhabit our ecosystems must be broken and that we must have lost the things we once shared. This is a distressing feeling, and not just because of the negative impact this perspective has on nature. But if that really were true, then we would be the only rational species living on a planet full of dimwitted, helpless creatures. Creatures that surround us every day. Dogs and cats, birds and squirrels, butterflies and flies—all less intelligent and therefore condemned by us to endless oppression or perhaps even extinction. That worldview is enough to make anyone feel excluded.

We are, as we always have been, part of a larger community. We are equipped with outstanding sensory organs, which allow us to understand and make the most of the environment where we belong. These senses also make us aware of other species, with all their sensory capabilities, and thus strengthen our sense of empathy and consideration for them. The ancient tie that binds us to nature is not and never has been severed. We have just ignored it for a while. And when we feel that we belong fully and completely to nature, conservation measures can be seen in a completely different light.

We are not just protecting nature somewhere out there or giving things up simply to prevent the extinction of apparently unimportant beetles or species of birds. On the contrary, with every step we take to help conserve the ecosystem that is the Earth, we are at the same time protecting ourselves and our quality of life, simply because we are a fully functioning part of the whole. Environmental conservation is and must be—literally and in the best sense of the word—about just one thing: self-care.

9

IN CLOSE CONTACT WITH TREES

WHY CAN'T WE communicate with trees the same way we communicate with, say, elephants? I like to compare the two because they have much in common. Both live in social groups and look after not only their young but also their elders. That famous elephant memory is also found in trees, and both communicate in languages that we didn't even recognize at first. Trees communicate through their interconnected root systems, and elephants communicate using sounds below the range at which we can hear, which they pick up through their feet over distances of many miles. Both engender in us a feeling of admiration and a desire to interact with them. We get a feeling of well-being when we run our fingers over the rough skin of both creatures, and what we would love above all is to get a reaction from them.

And here there is a distinct difference between elephants and trees. The animal lets us know what it likes. It responds by reaching out its trunk and making contact. There is definitely non-verbal communication going on here. And that is exactly what many nature lovers wish could happen with trees. But the little voice in my head, firmly grounded in the conservative views of science, immediately shouts: "No. That is taking things too far." And yet, I am a very inquisitive person, and when I consider everything that has been discovered in the field of natural sciences—take quantum physics, for example—then I'd like to take a closer look before I reject an idea as being impossible. And what has been discovered about trees sometimes leaves the little voice in my head lost for words.

CAN PEOPLE COMMUNICATE with trees? In order to answer this question, we first have to take a closer look at what we mean by "communicate." It is not enough that we consciously or subconsciously eavesdrop, so to speak, on the scents trees use to communicate among themselves. We have a physical reaction when we breathe them in, but for communication to happen, the trees also need to react to our signals. In the past, I have categorically refused to admit that this might be possible.

I AM GIVEN to neither religious nor mystical thinking. Perhaps this is a result of being unable to escape the obligatory trip to church every Sunday when I was a child. The preacher and the monotonous, unvarying rituals of the service bored me. I invented novel games to pass the time, such as pressing my eyes tightly shut while I gazed up at the overhead lights until kaleidoscopic patterns of light appeared on my eyelids—not exactly what my parents were hoping I would get out of this weekly visit. In high school, I greedily soaked up anything to do with science, as it seemed the only logical way to understand the world. And that's how it has been for me ever since, even though I know that in many cases scientific facts are simply the current best explanation we have for natural processes. These explanations are often revised and many statements are qualified. That's how science operates.

I would like very much to believe in higher powers. I think it would be emotionally rewarding and surely comforting. But I can't, and therefore I bring a healthy portion of skepticism to the table when non-scientists report experiences that sound unbelievable. I know that sounds strange coming from someone who writes about the feelings of trees and even talks about tree language, but these are both subjects on which even conservative scientists agree.

LET'S TAKE A moment to consider tree communication using the methods of modern science. Trees transpire chemical compounds. We are subconsciously aware of these compounds and we respond with changes in blood pressure. The tree, for its part, is unaware of our response—after all, we are not in contact with the tree in any way. And even if we hug the tree and talk of electric fields, which is one way we could mutually affect each other (because plants, like us, function partially by transmitting electric signals), there is still one huge obstacle: time. Trees, as we all know, are awfully slow. You can multiply the time it takes you to make contact with the tree by ten thousand to find out when you can expect a response.

If electrical impulses within the tree travel at a maximum speed of less than half an inch (1 centimeter) per second, and you make contact with the bark as you hug the tree, you could indeed get an answer right away.[31] At least you could if the signal is processed at the point of contact, but that is something we do not know. Certain processes are regulated in the roots—how much water the leaves can use, for example—and the distance from the canopy to the roots and back to the canopy (or to your hands) varies from tree to tree, but it is a long way. And now we are approaching one of the central questions about what it is to be a tree. Trees store memories, respond to attacks, and transfer sugar solution, and perhaps even memories, to their offspring. All these abilities suggest that they must also have a brain. But no one has yet found any such thing.

Many parts of a living tree, like most parts of its trunk, are not even active anymore. With the exception of the outermost growth ring, none of the interior is still in use. You could even say it was dead. Nothing happens here apart from a few purely physical reactions, which are the same as the reactions you see in wood after a tree has been cut down. There's the swelling and shrinking caused by getting wet or drying out, for example, as well as resistance to fungal decay thanks to tannins the tree stored earlier in its life that now act as a sort of waterproofing.

The vessels that transport water are found on the outermost growth rings. And that's why it's especially damp, even wet, there, which has the added advantage that most fungi can't grow in these rings. Fungi like damp conditions, but (with a few exceptions) if they get too wet, they drown. And because many species of fungi like to make life difficult for trees, it's a good idea for the tree to have a zone running around the outer part of its trunk that repels most attackers.

But, let's get back to the brain. Even in the outer parts of the trunk, most of the cells are dead wood. And so, we can confidently discard any speculation that important information is being processed here.

I'M USING THE term "brain" deliberately in this context, because I'm convinced that qualitatively valuable communication requires consciousness. If that were not the case, every computer would be a good communicator. After all, even a cheap electronic machine is capable of generating a response to your input. And this brings us to the question of whether plants possess something resembling consciousness.

Professor František Baluška at the University of Bonn has recently been looking into this. For some time now, he has been of the opinion that plants are intelligent—after all, they can process information and make decisions—but consciousness takes the discussion to a whole different level. If we could prove that plants have consciousness, we would have to radically change the way we interact with them, because we'd find ourselves facing the same kinds of issues that we face with factory farming in conventional agriculture.

Baluška, together with colleagues from around the world, including Professor Stefano Mancuso from the University of Florence, has come a little closer to answering the question about plant consciousness. To do this, Baluška and his colleagues sedated plants that feature moving parts, such as Venus flytraps. These plants catch their prey in a trap that snaps shut as soon as insects touch trigger hairs on the inner side of their double-lobed leaves. The two sides of the leaf fold together in a flash, capturing the insect between them, and the plant then digests its prey. The anesthetics the scientists used, which included some that are used on people, deactivated electric activity in the plants so that the traps no longer reacted when they were touched. Sedated peas showed similar changes in behavior. Their tendrils, which usually move in all directions as they slowly feel their way through their surroundings to find supporting structures to grow on, stopped searching and started to spiral on the spot. After the plants broke the narcotics down, they resumed their normal behavior.[32]

Did the plants wake up as we do when we come to after a general anesthetic? This is the critical question, because in order to wake up, you need one thing above all others: consciousness. And it was exactly this question that a reporter from the *New York Times* posed to Baluška. I really liked his answer: "No one can answer this because you cannot ask [the plants]."[33]

That was not the end of the story, however, as far as I was concerned, and to investigate further, I visited Baluška in Bonn. But before I get to our meeting, I'd like to first take a quick look at our shared history with plants.

LET'S CONSIDER THE question of two-sided communication from an evolutionary standpoint. In other words, for how long have humans and plants traveled together along the path of evolution and to what extent have they adapted to each other? We'll begin with trees, because they have a bit of a head start on us: the first trees elevated themselves above algae, moss, and herbs 380 million years ago. One of the reasons they did this was competition, which they could escape by growing taller. Plants that are able to unfurl their leaves above the leaves of others win the race for the sun. What could beat building an enormous trunk to lift your branches high over everyone else?

Obviously, this solution wasn't restricted to a single species. In the millions of years that followed, enormous forests sprang up, which had an unanticipated effect on other plants and animals. The trees breathed in vast amounts of carbon dioxide and bound it up, not only in their wood. Dead trees fell over, sank into swamps, and gradually formed coal.

So much carbon dioxide was removed from the air and so much oxygen was added that insects became larger than they had ever been before. The size insects can grow to is limited by the way they breathe. Oxygen is distributed through their bodies not by arteries and a circulatory system with a pump (the heart), but via a system of tiny tubes that delivers it directly to the cells. The longer these breathing tubes are, the less efficient they become, which means if insects are too large, they can't get enough oxygen. Based on the current levels of oxygen in the air, the largest insects can grow is about 6.5 inches (17 centimeters), and in the conditions we have today, no insects can grow any bigger than this.[34] Three hundred million years ago, however, trees were producing so much oxygen that the levels in the air were considerably higher than they are today. Instead of 21 percent, it was at 35 percent—and that meant insects could grow considerably larger, which is exactly what they did. Dragonflies with wingspans of 27 inches (70 centimeters) or more patrolled the skies and centipedes nearly 10 feet (3 meters) long scurried through the leaf litter on the ground.

THE CHANGING SIZE of insects is a great example of how animals adapt indirectly to trees, in this case to their metabolic processes. And here's a little aside. Like us, trees breathe in oxygen and use it up. After all, their cells burn sugar to get the energy they need to grow, just like our cells do, and this process requires oxygen. Producing sugar in the leaves through photosynthesis does indeed generate an excess of oxygen and store carbon dioxide, but if a tree does not also take in oxygen, it is not going to grow. Photosynthesis (producing food) and respiration (producing energy to process that food) are completely different processes. You can get a good sense of how little storing carbon dioxide and burning sugar have to do with each other if you observe trees in winter. At that time of year, trees burn the supplies of sugar they have gathered and stored over the summer, much as overwintering grizzly bears live off the layer of fat under their skin. Both trees and bears inhale oxygen with the air they breathe and exhale it out again bound as carbon dioxide. Because beeches, oaks, and other deciduous trees don't have any green leaves at this time of year, they are not generating any excess oxygen to release.

BUT, LET'S GO back to the prehistoric past. Long after the oxygen levels in the air had dropped once again and insects had shrunk to their present size, humans stepped into the picture. Fairly quickly, our ancestors learned to make fire and, for that, they needed wood. This, then, was their first important contact with trees. When exactly that contact happened is lost in the mists of time. We don't even know exactly when to start calling our ancestors humans. Even the date for the first appearance of modern humans—the species *Homo sapiens*—had to be revised in 2017. Until then, the date had been set in stone: we had been operating on this Earth for

200,000 years. Then, however, researchers in Morocco discovered even older remains. In caves at Jebel Irhoud, they unearthed bones and flint tools for making fire that dated back more than 300,000 years and indisputably belonged to modern humans.[35] For me, this was another good example of how scientific knowledge can change overnight.

The genus *Homo* first stepped into the spotlight 2 to 3 million years ago. Does this mean we should keep an eye out for adaptations in this time period? Why not even earlier? After all, even today we carry genes and skills around with us that we inherited from ancestors we would not begin to consider as human. And even if we consider the last 3 million years: is this relatively short time period long enough for trees to adapt to our presence? After all, for true communication or at least some kind of interaction to happen, we are not the only ones that would need to have adapted, and unfortunately, there is no evidence that trees have adapted to communicate with us. Trees have changed in recent times because of our actions, but the changes have little to do with communication. They are merely the result of people's efforts to select the most impressive specimens for gardens and urban beautification projects, forcing trees into a speeded-up version of evolution that corresponds to our desires.

IT HAS, HOWEVER, been proven that wild plants can make fundamental changes to adapt to the presence of people. A case in point is the bog orchid, which grows in the cool northern woods of North America with subspecies in Scandinavia and Russia. These orchids are on the lookout for pollinators for their white blooms, but in the swampy landscapes of the North there are few bees around. What there are in droves, as every vacationer can painfully attest, are mosquitoes. But these insects are not well known for their love of flowers. We have them pegged as nothing but annoying bloodsuckers.

And here is where the bog orchid comes in. It imitates the scent of humans to signal to mosquitoes that there might be a meal available. In search of victims, the insects happen upon the flowers and unintentionally pollinate them. But they don't fly away completely unrewarded. Even female mosquitoes drink more than just blood. Every once in a while, they enjoy a few carbohydrates on the side, which they get from a little sip of nectar.[36]

10

IN THE BEGINNING WAS FIRE

A COMPLETELY DIFFERENT WAY to get closer to uncovering a connection between people and trees is fire. Fire? That sounds a bit like having a discussion about animal husbandry over a steak. Firewood is basically tree bones sawed into small pieces so they can be burned. And yet fire is a good way to find out whether our interactions with wood (and therefore with trees) are rooted in some way in our genes. And there's strong evidence that this is indeed the case.

Does this sound familiar? You're in your own backyard or over at a friend's, someone lights a fire, and sooner or later everyone is standing around staring into the flames. This can happen even in summer on a mild evening when you don't need a fire to stay warm. Why do we do this? I'm sure we could both come up with a number of reasons: the atmosphere is romantic, we all love crackling logs and flickering flames. Here we are, stepping into the world of emotions, and emotions are the language of instinct. Our fascination with fire, therefore, is anchored in our subconscious. The only question is, are our positive emotions learned or inherited?

IF WE WANT to illuminate our relationship with forests and trees, it's crucial to begin with fire. It shows particularly clearly that our fate is inextricably bound to wood.

Fire is the most consequential achievement of humankind. Without blazing flames, our brains would never have developed to the size they are today. Cooking food makes many fruits—and meat—easier to digest, and this means that meals deliver significantly more energy. Also, as they lacked sharp canines or great physical strength, our ancestors had difficulty defending themselves against predators. Fire changed everything, because it's something all animals fear. The triumphant march of humankind around the globe would never have made it out of Africa had it not been for fire. How else would people have warmed themselves, for example, in the wide expanses of Siberia on bitterly cold winter nights?

Humans are not the only species that use fire. Predators are magically attracted by it—its bright glow drives prey species out from their cover, and sometimes their prey even gets burned, which means it's more difficult for them to flee. So-called firehawks in Australia have even been observed picking up burning sticks in their beaks and talons to spread wildfires so they can hunt the animals that flee from them.[37] In contrast to humans, however, eagles and their ilk can use fire, but they can never light it themselves. Even our closest relatives, chimpanzees, cannot start fires and, indeed, are clearly afraid of them. It is the ability to start fire that sets us apart from other species.

Just when the first flames were intentionally ignited is lost in the mists of time. However, a discovery in the Wonderwerk Cave in South Africa leaves no doubt that 1.7 million years ago humans were sitting around a fire they had made themselves.[38] Although there is no scientific consensus as to whether the first human-lit fires burned as far back as 4 million years ago, what is clear is that the warmth from wood has been part of our journey for a very long time.

A CHANGE OF scene. I was sitting around a fire with actor Barbara Wussow and meteorologist Sven Plöger. It was a small fire, because in the summer of 2018 the danger of wildfires in Germany was high thanks to a prolonged drought. We were filming for my television program. In each episode, I spent a night under the stars or, more specifically, under the canopy of the forest with two famous people. The fire sputtered, producing more smoke than heat. We had to get up close to cook over it. I held the handle of the pan with one hand while I stirred the food with the other. Not only did my clothing get thoroughly impregnated with smoke, but I breathed in a lot of it as well. And it occurred to me then that the affinity people have for smoke and smoked food might have its origin way back in time.

If our species has been around smoke every day for more than a million years, and if caves and later farm cottage kitchens with open hearths concentrated the smoke because of lack of ventilation, then maybe breathing in smoky air contributes to our instinctive sense of well-being? Just to be clear here: of course, no one likes to inhale smoke and then cough for minutes on end. And yet our lifestyles for the past few thousand years have left traces in our genes. If fire triggers an instinctive fascination that is possibly handed down to us in our genes, then why not a fascination for smoke as well or, better, for the smell of wood smoke?

Try testing this for yourself. Wood smoke or something that has been infused with wood smoke (it could be the clothes you were wearing the last time you had a campfire) smell good. At least they smell better than something like burned hair or, even worse, burned plastic. Burned hair triggers an instinctive physical alarm. Who knows, you might have inadvertently let your hair fall over the flames. Plastic and other synthetic substances, on the other hand, are so new that they barely register in our subconscious repertoire. In summary: most people find the smell of wood smoke pleasant; other things that burn, not so much.

It has not yet been proven whether delighting in the aroma of smoke or even being fascinated with fire itself is fixed in our genes. However, recent research has discovered changes in our genetic material that could be attributed to our relationship with fire. Professor Gary Perdew's team of researchers at Pennsylvania State University studied the genes of Neanderthals and the Neanderthal-like Denisovans, and compared them with those of modern humans. They discovered a difference that is significant when it comes to smoke. Smoke contains a wealth of cancer-causing materials, including polycyclic aromatic hydrocarbons (PAHS). PAHS are a result of incomplete combustion; in the human body, some of them are converted to other, similarly harmful, substances. People have been exposed to PAHS for 1.5 million years—and continue to be exposed to them today.

You, too, are breathing in PAHS all the time, although no longer around the campfire. Nowadays, most wood burning happens in woodstoves and home fireplaces. There are enormous numbers of these. In Germany alone, there are more than 12 million stoves and pellet heaters, and they burn more than half of all the wood cut down in that country. In the United States, an estimated 30 million people live in homes where wood is burned to provide heat.[39] Never before in human history have there been so many "campfires."

Smoke has an evolutionary impact. Then, as now, it was responsible for premature deaths, and over the course of hundreds of thousands of years it should have left a genetic imprint. And, indeed, it has. Perdew and his colleagues discovered a gene segment that distinguishes us from Neanderthals and Denisovans: the so-called Ah receptor. It inhibits the destructive effects of environmental pollutants, including those contained in smoke. Depending on the molecule, this receptor reduces toxicity for modern humans by a factor of one thousand.

What was the situation with Neanderthals? After all, the Neanderthal brain also benefited from food that was thoroughly broken down and made more digestible by fire. It was in some respects even larger than the brain of modern humans. But perhaps because of their smoky homes, Neanderthals didn't live as long as *Homo sapiens*. And it could be why, one day, they died out completely. At least that's what some researchers think, although these early humans probably developed other mechanisms to cope with fire and smoke.[40]

What interests me in this research is not why Neanderthals died out—when you look more closely, that's not actually the case. After all, quite a few Neanderthal genes are found in us, so they could simply have been integrated into the modern population. What I find interesting is that the results of this research get us closer to discovering that fire/wood/trees are so deeply rooted in our genes that, to this day, we can find traces of the connection. Our desensitization to smoke lingers still. This is the only way to explain why cigarette smoke doesn't wreak even more havoc on our bodies.

WHATEVER TIE THERE is between humans and trees, it doesn't always have to lead in a positive direction. One particular aspect of the relationship, wildfire, could hardly be described as a benign way of interacting with nature. Fire helped our ancestors clear large areas quickly without having to expend too much effort. Even today this method is used to free up large swaths of land for agriculture in Asia and Latin America. There is also no question that in the 1.5 million or more years that our ancestors have been lighting fires, countless huge forest fires have been started unintentionally.

You often read reports that describe forest fires as a natural occurrence. There may be some truth to this; however, the huge fires that are occurring with increasing frequency in western North America and Russia are certainly not natural as far as those ecosystems are concerned. The deciduous forests native to Germany hardly ever catch fire from natural causes; the coniferous forests of northern Europe, in contrast, do catch fire more easily. The trunks, needles, and bark of conifers are full of resin and other flammable substances. In dry summers, these trees are like gas tanks waiting to explode. But what tree likes to burn? Intact coniferous forests store a great deal of water in mosses, lichens, dead wood, and humus. All those soggy materials inhibit fire. Moreover, the lightning strikes that are the leading natural cause of forest fires are usually accompanied by heavy rains that extinguish the fires just as they are beginning to burn. When a fire starts in dry weather, it almost always was and almost always is human caused.

EVEN IF FIRE—INCLUDING, indirectly, firewood—has left its imprint in our genes, if we are looking for a tie between humans and nature, this connection is a bit too thin. It would be nice to find stronger support for our relationship. And therefore, in the next chapter, I will approach a slightly more shocking subject, literally: electric fields. Up until now, I always avoided this topic, because the whole thing seemed to verge a bit too far into the intangible and unprovable realm of esoteric thinking. Electric fields around trees conjured up images of the mysterious auras that are often evoked when people talk about trees sending us clear messages, speaking to us directly, and transferring their energy to us—if they are in the mood to do so.

However, since becoming familiar with the latest university research on the subject, I've looked at electric fields in a completely different light. I'm sure you will, too.

11

ELECTRIC TREES

SPIDERS COULD HOLD some useful clues to understanding more about the electric fields that surround trees. At least, that's what Dr. Erica Morley, a biologist at the University of Bristol, clearly thought. She was researching a method of aerial locomotion used by spiders called ballooning, where a spider shoots a long thread of silk into the air and lets itself be carried away on it. Unsurprisingly, this works particularly well with small, light juvenile spiders, which gather in large numbers at the end of summer. Their silvery threads fill the air on mild afternoons, leading to the term "old women's summer" in Germany because the threads are reminiscent of the long silvery-gray hair of older women (or so people used to think).

So, if you are a spider, how does a thread help you fly? It's really quite easy. The breeze catches the feather-light silk and, along with it, the spider dangling from the end. At least, that was the theory up until now. Morley, however, discovered that when a spider initially launches itself from a branch or leaf, other forces must come into play. The silken thread the spider shoots from its rear end lengthens quickly, and if it is not swept away correspondingly quickly, the spider could get tangled up in it. But spiders don't fly in high winds.

Some scientists suggest thermal processes are involved. For example, on warm days, the sun heats air on the ground, causing it to rise. But mass ballooning happens even on rainy days, when thermal air currents are hardly a factor. Moreover, spiders need to ascend rapidly when they leave their branches or they would fall to the ground before a gentle breeze could catch them.

The solution, it turns out, is electrostatic charges strong enough to move small objects. Imagine an athletic shirt made out of synthetic fabric. When you pull it off over your head, it sometimes crackles. In the dark, you might even be able to see tiny flashes of light, and if you look quickly in the mirror, you'll see some of your hair standing straight up. If electric fields are involved and actively employed, then the spiders must feel them and be able to align themselves with them. And what could be more helpful here than hair? This is exactly what spiders rely on, as Morley discovered.

For her experiment, Morley placed the little creatures into a polycarbonate box with aluminum plates at the top and bottom. The spiders were placed on a non-conductive cardboard strip in the middle of the box. The lower plate was grounded and the upper plate was charged to create a difference in the voltage between the bottom and the top. Through the hairs on their body, which, like ours, stand up straight in the presence of electric fields, the spiders registered the increasing voltage differential. They immediately lifted their rear ends, extruded a thread of spider silk, and launched themselves into the air.[41]

Electrostatic forces are certainly not the only forces that allow spiders to fly. The wind also plays an important role. But in order to launch from trees, especially on mostly calm days, this phenomenon is indeed of utmost importance, as the spiders in the box with their ability to react to electric fields clearly demonstrated.

THE EXPERIMENT LEADS to another question: How are electric fields generated around trees? Does this mean the mysterious aura that supposedly helps us communicate with trees exists after all?

The explanation is both simple and complex. The cause of the aforementioned forces are electric processes in the atmosphere. We have known about them for at least two hundred years. The ionosphere, a layer that begins 50 miles (80 kilometers) above the ground, is positively charged, whereas the surface of the Earth is negatively charged. The difference between the two is more than 200,000 volts. The differential increases the farther you are above the ground, and it does so quite rapidly. The difference in the first few feet above the ground in good weather is from 100 to 300 volts—every 3 feet (1 meter).[42] Under a thundercloud, the difference can be as much as thousands of volts per meter.

Despite this, you don't have a higher voltage around your head than you do around your feet, because your body is an efficient conductor of electricity. At some time in your life, you have probably felt a brief shock as you touched a car or a piece of plastic garden furniture. What's happening here is that your body is equalizing the voltage between the object you just made contact with and the ground. This way, you remain unelectrified and still have the same amount of voltage in you as the ground. This also means that the air around your head carries a higher electrical charge than you do, because air is a poor conductor and retains electricity for a long time. When it comes to trees, this differential is even higher thanks to their height. According to Morley, the difference between the tips of the branches of an oak tree and the surrounding air can be more than 2,000 volts per meter, and this difference can sometimes cause the tree to glow.

NOW IT'S TIME to return to possible interactions between people and trees on an electric level. Could it be that we react to these electric fields in the same way we react, for example, to changes in the weather? After all, it has already been scientifically proven that animals are not only aware of these fields but also actively use them.

A team of researchers led by Dr. Dominic Clarke (also at the University of Bristol) investigated bumblebees. When bees are searching for flowers, these insects orient themselves using a variety of different markers, including color, form, and scent. This potpourri for the senses is in itself a kind of communication—after all, flowering plants go to all this trouble only because they want to advertise that their nectar is available (in exchange for bees flying away with a dusting of pollen). Up until now, scientists researching these signals have focused on those things that people can also easily pick up on—first and foremost, what they can see, smell, and taste. Yet, at least as far as the bumblebees are concerned, this is not the complete picture.

Flowers, like trees, are surrounded by electric fields, which, thanks to the smaller size of the plants, are of course considerably weaker than the fields surrounding trees. Nevertheless, they are still strong enough that the tiny fliers are aware of them. What makes it especially easy for the bees to find the flowers is that the bees themselves are positively charged (because of the friction produced by their bodies when they fly), whereas the flowers are negatively charged. The difference in charge causes something else to happen: as soon as a bumblebee lands on a flower, the charge of the bumblebee (positive) and that of the flower (negative) cancel each other out. And that has a distinct advantage for other bumblebees. Normally, pollinated flowers change their color, shape, or smell, but that can take minutes or even hours. The change in electric charge, however, happens instantaneously and indicates to other bumblebees that there is nothing more to eat here.

To prove this, Clarke and his team constructed artificial flowers that could be electrically charged. The charged flowers offered a small sugar reward, whereas the neutral flowers contained a solution of bitter quinine. And lo and behold, the bumblebees visited the charged ones far more frequently.[43]

There are indications that honeybees use similar phenomena to transmit information. They, like the bumblebees, acquire an electric charge from friction when they fly. After they have foraged on flowers, they do their waggle dances for their audience of non-charged honeybees back at the hive, and the antennae of their insect audience react automatically to the differences in electric charge just as our hair does. And because antennae are sensory receptors, the bees are aware of the stimuli, which means it is possible that the dance could be an electric form of communication.[44]

IF WE TAKE a look around in the animal kingdom, we find other species capable of registering electric fields. Fish, for example, have a so-called lateral line organ in their skin that detects the Earth's magnetic field. The fish use this as a navigational aid, but that is not all. Sharks use differences in charges to identify prey. Just a few nanovolts—that is a billionth of a volt—is sufficient. (Remember, the difference at the branch tips of an oak tree can amount to many thousands of volts.)

And these discoveries bring us one step closer to the evolutionary story of humans—fish are vertebrates like we are, after all. We get even closer in the animal kingdom if we look at birds, which can also detect the Earth's magnetic field. Carrier pigeons, for example, navigate using these invisible lines. If you disturb the birds using electric fields, you affect their sense of direction, at least in the short term. From an evolutionary point of view, we are even more closely related to dolphins, which are thought to be as intelligent as great apes. In the twenty-first century, we learned that, like sharks, they react to differences in electric charges, probably also as an aid in locating prey.

NOW IT'S OUR turn. Why should we feel electric charges or, to put it the other way around, why shouldn't we? After all, our bodies are entities controlled by electric stimuli, and electricity flows through them constantly. Every piece of information that travels through our nervous systems, every thought that arises in our brains, is passed on by pulses of electricity, even though the amount of electricity involved—about a tenth of a volt—is minimal. That means, however, that we must be extremely sensitive to stronger currents, because a system designed for weak currents has to be easily disrupted using stronger ones. And with that we find ourselves in the middle of the disputed field of "electronic smog."

By now, it is beyond dispute that our body reacts to electric fields. According to the German Federal Office for Radiation Protection, the official safety limit set by European Union regulations is 5,000 volts per meter. This is also the limit set by the ICNIRP (International Commission on Non-Ionizing Radiation Protection).[45] That is above the peak value measured at the tips of tree branches, but there's more to it than that. Official limits are often set on the high side, which is why some countries are more stringent. In Latvia, for instance, the limit for apartment buildings has been set at 500 volts, while in Poland the limit is 1,000 volts. And here we are in the range of what can be found in nature.

The official limits apply to powerlines and permanent loads, and not to the peak levels in unusual weather conditions, which create electric charges in nature. The point here is not what the potential charge is at the end of tree branches, but whether we notice them, because if higher charges in the long term cause health problems, then the human body should be able to feel them.

And, at least at the cellular level, it turns out that it does, as a research team from the University of California discovered. They investigated skin cells, which react like electric sensors and orient themselves to weak electric fields when wounds are healing. In these fields, components of the liquid in the cells (polymers) arrange themselves along the cell wall, which is negatively charged.[46] So it seems clear that electric charges outside the body can disrupt internal processes. What is even more exciting, however, is whether we can also be aware of this. Here the evidence is anything but conclusive.

ONE LINE OF inquiry that leads in this direction is electromagnetic hypersensitivity. There are people who believe they can feel electromagnetic fields and that these fields adversely affect their health. According to the Federal Office for Radiation Protection, 2 percent of the population in Germany claim to be affected. There have been many studies on this issue, but no clear causal connections have been found. In addition, electromagnetic fields generate radiation that, depending on the source, travels along different frequencies. Some of these frequencies will be familiar to you—they carry signals for cell phones, radios, and televisions. There is no question that these signals affect electronic equipment, because that, after all, is the point. However, since the signals rarely travel directly to their goal—a radio transmission, for example, is more like a stone thrown into water that creates ripples in all directions—you and I are hit by innumerable radio broadcasts and transmissions every second. According to the authorities, these transmissions are so weak that not only do they do us no harm, but we don't even notice them. That, however, is debatable.

Don't worry. We are not going off on a wild goose chase here. Electromagnetic fields are closely related to electric charges and can also have a significant effect on electronic equipment. For example, if you get charged up as you walk across your synthetic carpet and then touch your computer, you can damage the electronic components in its circuit board. That's why the packaging on these components always tells you to ground yourself before handling them, for example by touching a metal radiator to rid yourself of any electric charges.

Electromagnetic radiation is also the reason people are worried about using cell phones. After all, you have to hold the phone up to your ear and, while it's there, it has to send out a signal that is strong enough to reach the nearest broadcast antenna. That is why the Federal Office for Radiation Protection here in Germany, which tends to be conservative in its judgments, recommends that you use a landline for calls if one is available. It also recommends keeping conversations on your cell phone as short as possible or texting instead—that way you don't need to hold the phone anywhere near your brain.[47] To me, all this sounds less than reassuring, especially as the suggestions don't address the impact of electromagnetic radiation on the nervous system: the basis for the limits is the warming of the adjacent tissue, a reaction more akin to how a microwave works. So, just to be clear, the question here is whether the transmissions from your cell phone are powerful enough to cook your brain or heat it up enough to damage it.

At the moment, there is nothing official to suggest these transmissions cause cancer, but clearly some important questions have been overlooked. If a system as finely tuned as our neural network works using weak electric signals, how might our internal data transmissions be affected by officially sanctioned gadgets (cell phones and the like) that create electric fields so strong that they heat up our brains? At this point, I don't want to go any further into what this all means for our health, either for people or for trees—for even with trees there is evidence of impacts even if the evidence is unclear. So, let's return to direct contact with trees and their electric fields.

I HAVE ALREADY mentioned that people can detect the discharge of electric energy. The brief crackle you hear when you touch a charged car door or plastic garden furniture, followed by a prick of pain, are evidence of your body's sensory abilities. The only question is how high does the voltage need to be for us to be aware of it and can training reduce this level?

For you to feel an electric shock, the difference between the charge you carry and the charge carried by the object you touch has to be at least 2,000 volts. This gets us into the ballpark of what has been measured at the tips of those oak branches. There is a problem here. To get to the ends of those branches, you need to climb the tree. To do this you would, of course, start on the ground—and then you would be carrying the same charge as the tree itself because you would be grounded at its roots. So, to experience the voltage in question, you'd need to start out like a bumblebee, that is to say, without touching the ground. One way you could do this would be to stand in a cherry picker on a rubber mat that insulates your feet from contact with the metal bucket. In that case, if the weather were exceptionally dry, which is when electrostatic charges are at their peak, it should be possible to receive a slight electric shock when you come into contact with the extremities of a tree. But this is not a situation you are likely to find yourself in.

TO ESTABLISH ANY kind of communication with a tree, you might be tempted to go out into the woods and hug a trunk to see if you feel something. But because both you and the tree are grounded in the earth, there is absolutely no difference in the charges you both carry, and so—nothing. Yet it is theoretically possible to experience the electric field surrounding a tree. This brings us back to hair. Your hair stands up on end when it is electrically charged because (unless it is really greasy or wet) it is a poor conductor of electricity. Therefore, you are grounded, but your hair isn't necessarily.

If the hair on spiders reacts to the electric fields of trees and their threads of spider silk, like hair, are repelled by the trees' electric charge, might something similar happen with us? Perhaps people with long hair could have the same experience with trees as they have when they pull a synthetic athletic shirt over their head or when they touch a balloon after it has been rubbed against something: when they get close, might their hair stand on end?

As I finish writing this book, it is unfortunately late winter, so this is not something I can try for myself. The trees are sleeping, and they have drastically reduced the water content in the parts of their bodies that are above ground. But, this summer, I will definitely start experimenting. Perhaps, though, I should send Miriam up that tree. After all, her hair is longer than mine...

12

THE HEARTBEAT OF TREES

WHEN YOU HUG a tree, nothing electric happens, because, as you now know, your voltages are the same. But might the tree be aware of your touch in some other way? There is one strong contender here—with young trees, at least—and this is a process known as thigmomorphogenesis, which is when plants grow more slowly after being touched. All you have to do, for example, is stroke your tomato plants for a few minutes each day and they slow their upward growth and put their energy into growing thicker stems instead.[48]

This, however, is not the plant saying it loves you, too, but rather the plant reacting to what it likely experiences as a breeze blowing by, because the wind elicits a similar response. The shorter the plant, the less leverage the wind gets and the less pressure there is on the plant's roots, so a tomato plant with a shorter, thicker stem is more stable. The same is true, of course, for movement caused when animals brush past plants—plants that are less stable are more likely to fall over. Therefore, it may well be that the way tomatoes or small trees respond to this kind of disturbance (not only from the wind) is part of their genetic repertoire.

If you've noticed that plants are healthier after you've stroked them, you're right. Scientists have discovered that plants stimulated by touch produce more jasmonic acid. This acid not only regulates height and triggers the growth of thicker stems so the plants are more stable, it also makes the plants more resistant to pests.[49]

IF YOU WERE hoping to hug a tree and get a hug back, this information must be disappointing. The responses I have described are simply a defensive strategy plants employ against external events they view as a threat. Moreover, if the tree is to experience your hug, it must be sensitive enough to touch that it can feel your arms around its trunk. A tree does indeed possess a certain sensitivity to touch, but in completely different circumstances. For example, if a neighboring tree or a metal post presses against its trunk, it will begin to grow around the obstacle. For this to happen, however, the pressure has to be strong and above all persistent over time—two conditions that are not met in a hug. Large trees in particular have thick bark as befits their stature, and nearly all the cells in the outer layers are dead, which means trees feel as much, or as little, with their bark as we feel with our hair.

We do, however, find a great deal of sensitivity in a completely different part of the tree: its roots. At this level, the tree works its way through the ground with its root tips, which contain brain-like structures. The root tips feel, taste, test, and decide where and how far the roots will travel. If there is a stone in the way, the sensitive tips notice and choose a different route. The sensitivity to touch that tree lovers are seeking is therefore to be found not in the trunk but underground. If it is possible to make contact, the roots would be the first place to try. They have the additional advantage that they are easy to reach and, in contrast to the upper parts of the tree, are active even in winter. However, they like neither pressure nor fresh air—and so there's no point exposing these tender structures because even ten minutes in the sun spells death for their tissue.

THE MOST RECENT scientific discoveries, however, offer something completely different: the heartbeat of trees. Heartbeat? Trees, of course, do not have hearts like we do, but they need something that performs a similar function or the most important processes in their bodies would not work.

What blood is to people, water is to trees. I have written a lot about how water is transported up into the crown of the tree; exactly how that happens has not yet been adequately explained. The most popular theory, that moisture is drawn to the uppermost twigs by transpiration, doesn't work. According to this theory, water evaporates out of the leaves and that creates a vacuum in the trunk. This vacuum then draws water up out of the roots and the surrounding soil. Unfortunately for this theory, water pressure in the trunk of deciduous trees is highest in early spring. At this time of year, there isn't a single green leaf on the tree and so nothing can transpire.

The other attempted explanations (osmosis, capillary action) don't work either, so we are currently without answers. Or, we were. Dr. András Zlinszky at the Balaton

Limnological Institute in Tihany, Hungary, is shedding some light on the matter. Some years ago, he and colleagues from Finland and Austria noticed that birch trees appear to rest at night. The scientists used lasers to measure trees on calm nights. They noticed the branches hung up to 4 inches (10 centimeters) lower, returning to their normal position when the sun rose. The researchers started talking about sleep behavior in trees.[50]

Clearly, Zlinszky could not get this discovery out of his head, and he decided he needed to investigate further. He and a colleague, Professor Anders Barfod, measured another twenty-two trees of different species. Once again, they documented the rise and fall of the branches, but this time some of the cycles were different. The branches changed position not only morning and night, but also every three to four hours. What could be the reason for this rhythm?

The scientists turned their attention to water transport. Was it conceivable that the trees were making pumping movements at these regular intervals? After all, other researchers had already determined that the diameter of a tree's trunk sometimes shrinks by about 0.002 inches (0.05 millimeters) before expanding again. Were the scientists on the trail of a heartbeat that used contractions to pump water gradually upward? A heartbeat so slow that no one had noticed it before? Zlinszky and Barfod suggested this as a plausible explanation for their observations, nudging trees one step further toward the animal kingdom.[51]

A heartbeat every three to four hours is, unfortunately, too slow for even the most sensitive person to feel when they hug a tree, and so, once again, we have failed in our search for a noticeable signal from the tree in response to our touch.

THERE IS ONE last possible way to connect with trees that I would like to look at in more detail: our voices. Our most important form of communication is verbal, and quite a few people try to talk to trees or to their houseplants. Why did I say "try"? This is something people actually do, and they expect the plants to react in some way. There are also wine growers who play a wide variety of music to their vines and believe that certain musical genres lead to better harvests.

Is there a grain of truth to this? Can plants even hear?

To that last question, I can answer without hesitation in the affirmative. This was tested years ago with *Arabidopsis*, a genus of rockcress beloved of scientists. Beloved because it grows well, it reproduces rapidly, and it's easy to keep track of its genes. Scientists discovered that the roots of *Arabidopsis* oriented themselves toward clicks in the frequency of 200 hertz and then grew in that direction. They can also produce sounds that function as a kind of Morse code.[52]

PROFESSOR MONICA GAGLIANO of the University of Western Australia discovered that pea roots can hear water flowing underground. To test this, she buried three pipes. In the first, she played a recorded swooshing sound, in the second real water flowed, and in the third there was the recorded sound of flowing water but no actual water. The plants were not fooled. Their roots grew only toward the real water. And when they were not thirsty, they showed no signs of movement. But is that really equivalent to hearing? Gagliano and her team thought that if it were, the roots would find white noise irritating, and this is exactly what they observed.[53]

So plants, including trees, can hear. Just like us, they use this skill for a specific purpose. We don't hear most low-frequency sounds because we don't need to, and plants also listen to those things that are important to them, such as water underground. But what about all those reports of vineyards filled with classical music to encourage bountiful harvests? What about the testimonials from people who talk to trees? From a sober, scientific perspective, the roots' auditory abilities are excluded from these responses, because roots grow in the relatively well-soundproofed realm beneath the surface. Therefore, we need to look above ground. Let's look more closely at stems, branches, and leaves. Is there any evidence of acoustic responses in these?

OVER A PERIOD of many days, a television team from West German Broadcasting in Aachen offered sunflowers at the Jülich Research Center a range of different sounds, including classical music. They discovered no growth differences at all between the plants.[54] Does that mean there is nothing going on acoustically above ground? Let's not give up so quickly. Perhaps music is the wrong starting point. We'd do better if we looked for sounds that mean something to plants.

How about, for example, the nibbling of caterpillars, an ominous sound to plants of all species? At the University of Missouri, this was exactly what they investigated. Researchers put caterpillars on samples of *Arabidopsis*. The vibrations caused by the caterpillars munching were enough to shake the plants' stems, and the researchers used laser beams to record the vibrations. When researchers then played these vibrations to plants that were not being eaten, they produced particularly large quantities of defensive chemicals when they were later attacked. Wind and other sounds with the same frequency did not elicit a reaction from the plants.[55]

Arabidopsis, then, can hear, and this makes perfect sense. Thanks to acoustic warnings, it is able to recognize danger some distance away, so it can make appropriate preparations to defend itself. What is particularly important here is that the plants ignore noises that pose no threat to them. These noises probably include human voices as well as different styles of music. What a shame. The reports of crops that can clearly distinguish classical music from rock music sound so delightful.

We still need to monitor one thing, however: music with notes that approximate the sound of munching caterpillars. That could happen, and, if it did, an acoustic misunderstanding might prevent the plants from enjoying Mozart.

I CAN WELL understand people's desire to communicate with trees. To sit under these giants, run your hands over their bark, and feel secure—all this would be even more special if there were an active, positive response to your presence or, even better, to your touch. I am not going to deny that something like that might be possible, but conservative science at least has no proof that it could happen. And even if this were the last word on the subject, does the tree have to respond? Could it not be that people and trees live in completely different worlds? After all, our species has existed for only 0.1 percent of the time that trees have been around.

Although trees clearly feel nothing of all this, we, for our part, definitely experience a physical reaction, one that I will look at in more detail later. For the time being, it should be enough that we feel good around trees—and I hope we can then be content to allow them to live their own wild lives.

13

WHEN EARTHWORMS TRAVEL

IF WE LOOK back many thousands of years, we would certainly not see any foresters among our ancestors. Back then, no one was interested in forestry, and yet people have always had a huge impact on what kinds of forests are out there. Mostly indirectly, it must be said, and through the world of animals. Before I turn to the smaller creatures such as earthworms (yes, earthworms change forests!), I would like to spend some time with mammals.

IN MANY CASES, trees can do just fine without any mammals. For hundreds of millions of years, these furry creatures were not even in the picture. These species began their triumphal march after the dinosaurs died out 66 million years ago and then evolved into large herbivores that munched their way through the trees' leaves and branches. Since then, trees and mammals, including our own species, have interacted. And because our ancestors cut down trees for firewood, the shared story of people and trees is also an ancient one. That is significant because the forests of the Northern Hemisphere are relatively young, dating back to the last ice age or, it would be more accurate to say, to the end of the last glacial period.

In ice ages, the polar caps are frozen, and as they are still frozen today this means that technically we are still living in an ice age. Fluctuations within an ice age are called glacial periods (when glaciers advance) and interglacial periods (when glaciers recede). The last glacial period ended about ten thousand years ago, depending on exactly what region of the Northern Hemisphere you are talking about.

Where glaciers many miles thick scoured the ground, no higher life-forms could survive. When the ice masses melted, they left behind sand and coarse gravel, and the land had to slowly green up once again. Even in those areas not covered by ice, temperatures remained so low and winters lasted for so long that it was impossible for trees to grow. Woolly mammoths and reindeer survived on lichens, mosses, and dwarf shrubs.

Once the ice was gone, the trees returned from their refuges to the south. Not the individual trees that had escaped the ice age, of course, but the next generations of the trees whose seeds birds had carried south, which was what allowed the populations to survive. Year by year, mile by mile, the ice melted. Forests followed the receding glaciers north and people arrived. Our ancestors had been around for tens of thousands of years, both during and shortly after the glacial period. They were not farmers but efficient hunters who mostly targeted large herbivores.

Although scientists don't agree whether people were solely responsible for the mass extinction of mammals during this glacial period, a great deal of evidence suggests our ancestors were at least a strong contributing factor. The mighty woolly mammoth disappeared from Europe ten thousand years ago, as did the woolly rhinoceros and innumerable other large herbivores. And the same thing happened in other parts of the world. The ice melted and humans appeared. In North America, not only mammoths but also wild horses and camels breathed their last.

Our ancestors truly were protectors of the returning oaks and beeches, because one of the favorite foods of herbivores is tree seedlings, a dietary preference that is exploited to this day to keep heathlands open. If nature were left to her own devices, forests would grow on heathlands, but European regulations make sure that doesn't happen—after all, the romantic image of the pre-industrial agrarian landscape must be maintained. And so, sheep are pastured on there, and they do the same thing as the wild horses and aurochs before them: they eat the seedlings of beeches and other trees, thus preventing reforestation. Forest suppression by herbivores might have happened over large swaths of the Northern Hemisphere if our ancestors had not had such an appetite for meat.

OTHER UNINTENTIONAL CHANGES to forests happened considerably later, also because of animals whose populations were affected by humans. In this case, however, we are not talking about hunting and reducing populations, but the exact opposite. The animals are considerably smaller and don't belong to the range of animals we consider food. I'm talking about earthworms.

Earthworms are held in high esteem these days. They improve soil by consuming dead organic matter, mostly leaves from trees and plant debris. Because they always get a bit of soil in their mouths when they eat, their guts extrude a fertile mix of earthworm poop and minerals. This crumbly soil stores water exceptionally well and provides habitat for numerous tiny organisms. Slime-coated earthworm tunnels aerate the soil and ensure that even in storms, water can seep into the ground. The earthworm really should be a mascot for home gardeners, and the presence of a multitude of these animal helpers is a sure sign of a well-tended vegetable plot.

The situation is very different in parts of North America, where earthworms damage large areas of forest and endanger many plant and animals species. How can that be?

Northern forests lost their earthworms in the last glacial period, and their ecosystems have completely adapted to the absence of these soil-dwelling animals. One sign of this is the spongy layer of half-rotted leaves on the forest floor, filled with busy bacteria, fungi, and mites adapted to these conditions. The wiggling invaders from Europe gobble their way through this thick layer, leaving the forest floor bare and destroying the foundation of life for these tiny ground dwellers. Not only that, but plants that depend on this "forest compost" also disappear. In addition to consuming dead material, earthworms eat seedlings and seeds. What changes will happen in these forests in the long term is not yet clear, because these invaders are just getting going, but changes are inevitable given such massive alterations to life on the ground.

How does an earthworm travel? Very easily. Hundreds of years ago, earthworms arrived in the soil clinging to the roots of plants settlers brought with them to their new homes. The earthworms themselves didn't need to make the journey, just their remarkably resilient eggs. These days, anglers also play their part because earthworms are popular as live bait. Any that are not used are simply discarded near the shore. In the northern part of North America, any earthworms that are released have an easy time, because there are no native earthworms and therefore no territorial battles to be fought.

The problem of invasive earthworms exists on every continent, of course. Think, for example, of the cheap houseplants for sale in supermarkets—earthworms can travel in comfort in their pots from China to Europe, to mention just one possible journey. In places where earthworms already occupy the ecosystem, it is clearly more difficult for the new arrivals to get established and set massive changes in motion. But the more disturbed the landscape and the more woodland that has been turned over to agriculture, the easier it is for invasive species to spread.[56]

FUNGAL SPORES ARE much tinier than earthworms. As a general rule, the smaller alien species are, the more easily they are imported—a process that has existed for as long as people have set out on journeys. The tiniest particles literally cling to the heels of globetrotters and so arrive in places they would never otherwise have settled. Take, for example, a fungus that originally made its home in Korea. Whether in exported goods or on the shoes of trekking tourists, the spores of this fungus managed to get as far as the North Island of New Zealand.

Here, in the Waipoua Forest, formidable conifers grow. These are the ancient and mighty kauri trees. The largest of these, *Tāne Mahuta*, is almost 15 feet (4.5 meters) wide. And its age is as impressive as its girth. It has been standing in the forest for at least two thousand years, but how much longer it will survive is unclear. The imported Korean fungus is starting to extinguish life from the New Zealand giant. It destroys the roots, which eventually will kill the whole tree. Unfortunately, there are no "tree doctors" who can help.

Everything looked so promising at first. White settlers had left only a few kauri forests standing, and in the twentieth century, these were protected when Europeans finally stopped using both the kauri tree's wood and its valuable resin.

The original inhabitants of New Zealand, the Maori, had traditionally gathered resin from the roots of dead kauris, which contained enough nuggets for their needs. They processed it into chewing gum or dyes for tattooing. But the amount of resin in the roots of dead trees was not sufficient to satisfy the settlers' economic demands, and they tapped into living trees to collect resin to make lacquer and glue for ship building. Before the invention of synthetic dyes and petroleum-based substitutes for turpentine, resin was the go-to raw material and this natural product was in high demand. The tapped trees weakened. Added to that, shipyards and sawmills were booming, which meant that one giant kauri after another was brought to the ground. Eventually, the last kauri forests were protected, at least from commercial interests.

But then came the imported fungus with the awkward Latin name *Phytophthora taxon agathis*, commonly known as kauri dieback. It wasn't until 2008 that this organism was identified as the culprit attacking trees and making them wilt. One particular feature of kauri dieback is that it spreads along hiking trails, which often pass directly over the trees' roots. The situation is exacerbated when communities try to boost tourism by building more hiking trails and mountain-bike runs.

In the face of these pressures, the efforts of rangers to get hikers to clean their boots seems almost naïve. Visitors often walk right past the cleaning stations without even noticing them, and even when they are used, boot cleaning does little to solve the problem. Fungal spores measure between 0.00012 inches and 0.008 inches (0.003 and 0.2 millimeters), which makes them about the size of a speck of dust. How can tourists, eager to hit the trail at last, clean their boots thoroughly enough that they track absolutely nothing with them under the valuable trees? The only effective measure would be to keep visitors out of the last kauri forests, but authorities in cities such as Auckland have declined to do this—they are afraid of losing tourist dollars.

MEANWHILE, THERE ARE similar reports from other places around the globe; only the species of fungus and the trees change. Right now, in Europe, it's ash trees that are threatened. The culprit is known in Germany as *Falsches Weisses Stängelbecherchen*, which translates as "false little white stem cups," a reference to the shape of the fungus' fruiting bodies. In English-speaking parts of the world, it is known as ash dieback. The spores were imported to Europe with global trade with eastern Asia, and the fungus has been infecting the common ash since the turn of this century. It makes its way via the leaves to new growth and then into the wood. As infected tissue dies back, first thin branches and then whole sections of the crown wither and die. The disease was first identified in the United Kingdom in 2012, and by 2019, infections had been confirmed across 80 percent of Wales, 68 percent of England, 32 percent of Northern Ireland, and 20 percent of Scotland.[57] So far, the disease has not been reported in the United States.[58]

When they notice trees dying, many foresters begin frantically cutting down infected trees. However, felling trees doesn't help combat the disease, as there is no way it can be stopped. The fruiting bodies of the fungus grow on the central veins of last year's discarded leaves. From there, the little cups release their spores into the surroundings, hoping they will land on the fresh green leaves unfurling on the trees.

Cutting down trees is the desperate attempt to salvage the wood while it's still worth something. Some of my colleagues in Germany are writing off whole stands of ash and allowing perfectly healthy trees to be harvested. And with that they are hastening the decline of the species. In every stand, you can find a small percentage of healthy or only mildly infected trees. If you were to leave these more robust trees standing, then they could reproduce and their offspring could create healthy woods again. And yet concerns about revenues, combined with the attacks of the aggressive newcomer, mean more and more ash are disappearing.

The one-two punch of money and fungus is leading to an outcome similar to the one we are seeing in New Zealand: the character of the forest is changing. The changes are affecting not only the trees, but also the plant and animal communities that are losing the basic necessities they need for life. And so, for example, the little white stem cup, which looks almost exactly the same as the false little white stem cup that causes ash dieback—you can only tell them apart by looking at their spores—is now disappearing. The ash bark beetle is also on the losing side, although it doesn't know it yet. Like all bark beetles, it attacks weakened trees exclusively, and there are plenty of them to go around at the moment. However, once most stands of ash have disappeared, its fate will also be sealed.

I HAVE TO say that reports like these give me a guilty conscience. When I travel to forests far away—even if I'm traveling so I can lend my support to local conservationists—am I not providing a means of transportation for small disruptive organisms? I wear the same hiking boots no matter where I am, and when I return home, my boots are definitely not absolutely and completely clean. And it's not just fungi that are a concern. They are just one example of countless tiny organisms that you cannot see with your naked eye but are crucial if larger ecosystems are to function.

New York's Central Park shows us just how little we know. A team from Colorado State University led by Dr. Kelly Ramirez took soil samples every 50 meters and examined them for bacteria and similar small beings. As it would have been almost impossible to tell all the species apart under a microscope, the team did genetic analyses. To their surprise, they found 122,081 species of bacteria alone, most of them not yet known to science.[59]

Okay, they are "only" bacteria. But the significance of species in an ecosystem increases the smaller they are. The smallest creatures in the soil are the first link in the food chain, comparable to plankton in the ocean. If a large number of them have not yet been discovered, let alone researched, then you can imagine how little we really know about ecosystems.

Unfortunately, these bacteria are really small and stick to shoes even better than fungal spores do. As goods are traded and people spend their vacations in distant places, species are distributed in all directions across the globe. Wherever they land, the cards of life are reshuffled.

Before you begin to feel too bad: nature does the same thing. Long-distance travel? Yes, we are not the only ones to have airlines. Animals have air carriers, too. In the animal world, migratory birds cover great distances and, naturally, they don't wipe their feet before they take off. And they indulge in another behavior that allows even larger organisms to take long voyages: dust baths. Birds love to flap around and stir up dust and humus, so these fine particles get into their feathers. Dust baths help birds rid themselves of parasites, which are shaken out along with the dirt. Inevitably, some ground-dwelling creatures remain on board: not only fungal spores and bacteria, but also creatures such as springtails, so-called because they can catapult themselves into the air using their tails. There can be more than 100,000 springtails crawling around in every 10 square feet (1 square meter) of forest floor. In addition, there are hosts of other species, such as beetle mites or bristle worms. A few of these guys could certainly wander into a bird's plumage by mistake. They go along for the ride as the birds migrate, and when the birds get to their destination, they offload their passengers when they take their next dust bath.

SOMETIMES AIRBORNE TRAVEL fills in gaps in ecosystems, as I experienced in my own forest. University students were surveying one of the older spruce plantations. They wanted to see how species composition changes over time. There are native springtails that specialize in beech trees and really don't like spruce, so the students were not expecting to find any. And indeed, they didn't, but what they did find were species that are adapted to conifers and therefore cannot be native to the area—the only trees that grow here naturally are deciduous ones. The only explanation for the presence of non-native springtails is airborne travel on birds.

Birds can even fly fish to new places. Once as part of my training to become a forester, I visited a pumped-storage hydropower facility in the Black Forest. It consisted of an artificial lake from which water could be released into the valley below. When the water flowed down through pipes, it turned turbines to generate power. The water was released whenever the demand for electricity from the grid spiked. When there was excess capacity in the grid, some of the power was used to pump the water back up to the lake, where it was stored once again. According to the docent who was leading the tour, the lake was emptied and cleaned every once in a while, revealing tons of fish. But where on earth had they come from? The answer is quite simple: their eggs had traveled tucked in the feathers of ducks that discovered the lake and decided to stay a while, offloading the stowaways.

SO IT IS not only people who introduce alien species. The difference between us and our fellow creatures is, of course, that in these days of global trade and travel, we have speeded up the rate of introductions so much that nature cannot adapt quickly enough. The key words here are global trade. Because this trade increased dramatically after Christopher Columbus' voyage to the Americas in 1492, species that arrived after this date are considered to be non-native.

According to the German Federal Agency for Nature Conservation, almost three thousand alien species of plants, animals, and fungi have established themselves in Germany since then.[60] These include crops that were imported intentionally, such as potatoes, sweet corn, or pumpkins. Those crops don't have a role to play in nature because they would not survive if they were not being actively farmed, but that is not the case for about eight hundred other imported species. Raccoons, raccoon dogs, and chipmunks are three mostly harmless examples from the animal realm. I've already mentioned the fungal spores that travel on shoes and the abandoned earthworms that have far more detrimental effects on the forest. On top of all that, we have animals moving into new habitats. Even though the animals are doing this of their own accord, they are moving only because we have changed the landscape so drastically. Red wood ants, which belong in northerly regions or at high elevations, have only become established in lowland areas in Germany because that is where the coniferous plantations are. Here you can also find crossbills, which specialize in extracting the seeds from pine cones and therefore are definitely not fans of Germany's native beech woods.

Invasive species have wreaked and continue to wreak havoc in the United States, as well. The mighty chestnuts that dominated eastern forests were wiped out by a fungus that arrived from Asia a century ago. American chestnuts had survived for millions of years and were destroyed in forty.

The trees die back to the ground. They can put up shoots, but the new growth soon dies back again. Today, scientists are trying to restore American chestnuts by breeding crosses between American chestnuts and Chinese chestnuts that are immune to the fungus. Interestingly enough, this is possible because more than 60 million years ago, Asia and North America were joined together in a super-continent and so many of the trees growing on these two continents today originally grew together. After the continents split apart, the trees on the separated landmasses took their own evolutionary paths. Invasive insects are a problem, too. Since the early 2000s, emerald ash borers, also from Asia, are threatening to do to North American ash trees what ash dieback is currently achieving in European forests: kill large swaths of trees.

ECOSYSTEMS IN EUROPE and around the world are currently experiencing a chaotic mixing of species, and no one knows what the outcomes will be. Some kind of new natural order will eventually play out as soon as the involuntary global animal and vegetable tourism stops. But we can't predict what this new order will look like.

The way we spread fungi and bacteria around shows that we do not yet take the consequences of global trade and tourism seriously enough. Sometimes I feel we do not respect nature the way people used to do. There was a time when trees and nature in general played a much more important role in our lives.

14

THE SACRED TREE

CLOSE TO THE forest I manage, a hill called the Michelsberg rises up out of the landscape. Its green dome is topped by a small white chapel. The chapel is not only a striking feature in the landscape, but it also marks the spot where pagan customs ended in this part of Germany. There were once trees on this hill under which people made animal sacrifices. The practice existed for a long time, a very long time.

One day, while I was trudging along among the ancient beeches, I noticed a man seemed to be looking for something. I am always alone in the Eifel forest on weekdays as I do my job as a forester, so I noticed the man right away. Was he in need of help or had he just lost his way? I went over to him and, as things turned out, I'm glad that I did. The man was employed by the office responsible for the preservation of historic sites, and he told me a few things about the history of my forest. That day, he was looking for traces of an old Roman road that had been reclaimed by the forest. He pointed to faint indentations, about two thousand years old, visible under the layer of leaves.

So, the ancient forest road known as the Roman way really did date back to Roman times, then? No, the researcher told me, it's a bit more complicated than that. Although Romans had indeed used the road, it had originally been made by others. The road led to a sacrificial hill, the Michelsberg. It was laid down about ten thousand years ago by people who revered the domed summit with its stand of ancient trees. It was only when Christianity triumphed over the local ancient religion that people changed the way they used the hill.

As in all the other places where people were converted to Christianity, the priests had the trees chopped down and built a chapel where the people used to make their sacrifices. From then on, anyone who sought out the ancient site to follow the old religious rites had no choice but to make the pilgrimage to a Christian place of worship as well. The last sacrificial fire was lit around the year 800 CE, and when it went out, the old religion was extinguished, too. Or was it?

Remnant rituals from age-old forms of tree worship are practiced to this day. Take the southern Italian region of Basilicata. There is a form of tree worship there that must be thousands of years old, at least, and possibly dates back to the Stone Age. Although it was outlawed around the year 725 CE, it continued to be practiced and became integrated into rapidly spreading Christian rituals. At its center stands the marriage of trees. It is not a marriage in the traditional sense, because after a complicated ceremony, the trees are cut down. On the Sunday after Easter, a troop of people well versed in the rituals marches out into the woods to find the bridegroom. He must be an oak that has grown tall and straight. That day, the tree is merely flagged. A week later, people go looking elsewhere for the bride. She must be an evergreen, so a conifer or perhaps a holly. Her beauty is judged by her splendid, evenly proportioned crown. The trees breathe their last on Ascension Day, because that is the day they are both cut down. Oxen pull the trees into the village, where they are married on the holy day of the Pentecost. The ceremony involves attaching the crown of the holly to the trunk of the oak so they look like a single tree. Everything happens slowly, according to strict rules, with the lively participation of the local population (and today sometimes tourists, as well).[61]

OTHER REGIONS HAVE the same kinds of rituals with similar roots. In many places, a May tree is put on display. Sometimes more than one. Where I live in the Eifel and throughout the area around Bonn, there are hundreds of them. The recipients of these trees are young women whose admirers head off into the forest the night before the first of May. There they steal a tree. The less bold among them buy their trees from foresters or local youth groups, which use the sale of trees as a fundraising activity. Those who do not pay find their prizes secretly carried off in the early hours of the morning by members of these groups.

The only suitable tree in this region is the birch, which has often not yet leafed out because the weather is still so cold. They are decorated with garlands of crepe paper and then carried to the loved one's home—although today this often means driving them over in the open trunk of a car.

I myself brought back a few birch trees for my Miriam, not always with the blessing of my future father-in-law. My friends and I would stumble around Sinzig's town forest, where by the light of a flashlight we chopped down a tree whose girth presented quite a challenge for our blunt axes, but we persevered. After all, I wanted the tree to be large, the largest in the neighborhood. And, indeed, it towered above the eavestrough on the second floor of the house. I'd neglected to consider that for the next few days the wind would batter the birch against the metal. Tying the tree to the downspout was also unfortunate. The resulting dents and my father-in-law's admonishing look made me feel guilty, but in May of the following year, all that was forgotten and the game was once again afoot.

What deeper meaning lurked behind our actions? We young guys wanted to give our girlfriends a token of our affection that could be seen by the whole neighborhood. An ancient idea buried deep in this tradition lives on in the modern custom. There's a reason the tree is put on display in what we often refer to as the merry month of May. The name comes from the Romans, who dedicated the month to Maia, the goddess of fruitfulness—this is the original significance of the May tree.

In later years, I injected the occasional frisson of danger into this custom. One of the reasons it was so exciting when I was young was that we did not want to be caught by the forester. Then for many years I was that frightful figure on the lookout for young men as I patrolled my forest in the days leading up to the first of May. Birches are an important species in German forests because they are the first to grow in open disturbed areas. I had them in my forest because hurricanes Vivian and Wiebke uprooted thousands of spruce trees in 1990. The first trees to take root on the open ground were birches. At the time, many of my colleagues regarded birches as weed trees, but I was happy to see trees coming back into my forest for free. Even back then, I had little time for conifers. Fifteen years later, many beautiful stands of birch stood rustling in the breeze. They were perfect. Perfect to be used as May trees. Of course, I remembered my own youth and came up with this compromise: I didn't police the young men from the villages around my forest. They could hunt for birch trees by night undisturbed by me. The neighboring villages, however, had to ask for permission and buy a permit for the modest sum of about €10 (US$10).

A GLANCE AT other countries shows that Germany and Italy are not alone in their retention of ancient tree worship. Here's another example. In Cyprus, in front of the grotto of Saint Solomoni, stands a pistachio tree whose branches are hung with pieces of cloth. People believe that the knotted rags help cure diseases of the eye.[62]

In Celtic-speaking countries—Ireland, Scotland, and Cornwall—there are what are known as Clootie Wells. These are springs with a tree growing nearby. As in Cyprus, strips of cloth are tied to the tree. These ceremonies are said to help people recover from illnesses,[63] which is why the trees were sometimes also called "wish trees."

Easter fires are also widespread in Christendom. They are relicts of German religions that included these fires as part of their worship of trees. When areas were converted, these rites became integrated into Christian celebrations leading up to Easter. The fires you see in modern times are the rekindled flames of ancient Germanic religions that worshipped nature.

TODAY, MANY PEOPLE feel a strong pull toward a deeper connection with nature, perhaps even to religions that worship nature, and yet in recent decades reason and science have held sway. You can see this in the dwindling congregations of mainstream churches. To get closer to nature in times like this, it can help if we step into the realm of philosophy. And because I know next to nothing about this field of inquiry, it seemed like a good idea to engage in conversations with experts.

15

THE DISAPPEARING BOUNDARY BETWEEN ANIMALS AND PLANTS

In 2018, a German daily newspaper, the *Süddeutsche Zeitung*, asked me if I would be interested in having a conversation with the philosopher Emanuele Coccia, who had just written a book about plants. Because I enjoying exchanging ideas with scientists in many different fields, I was happy to say yes. It turns out it was a good decision, because Coccia gave me the opportunity to consider trees from an entirely different perspective. This perspective reinforced much of my thinking while pushing me to reflect more deeply in other areas.

His publisher sent me a copy of his book *Die Wurzeln der Welt* (published in English as *The Life of Plants*) so I could prepare myself for our conversation. The German title translates as "The Roots of the World," and the book really does cover this. It upends our view of the living world completely, putting plants at the top of the hierarchy with humans down at the bottom. I had been giving a great deal of thought to this myself. Ranking the natural world and scoring species according to their importance or their superiority seemed to me outdated. It distorts our view of nature and makes all the other species around us seem more primitive and somehow unfinished. For some time now, I have not been comfortable with viewing humans as the crown of creation, separating animals into higher and lower life-forms, and treating plants as something on the side, definitively banished to a lower level.

And so I found the conversation with Coccia most refreshing. He visited our Forest Academy accompanied by a journalist and a photographer from the newspaper. A small bearded man, Coccia turned up in a blue suit and blue checkered tie, completely inappropriate attire for the outdoors, even though we had agreed that we would take a walk in the forest together. Coccia is an unconventional thinker and something of a maverick, as he was clearly showing with his choice of clothes. Although he is from Italy and now teaches in France and writes in French, he also speaks fluent German because at one time he studied and worked in Freiburg.

After our first cup of coffee, we were soon deep into our main topic: trees and plants in general. Coccia argued that our biological classifications are not grounded in science. They are strongly influenced by theology and are dominated by two ideas: the supremacy of the human race and the world as a place humans must bend to their will. And then there is our centuries-old compulsion to categorize everything. When you combine these concepts, you get a ranking system that puts humankind at the top, animals in the middle, and plants way down at the bottom.

I listened, absolutely fascinated by what he had to say. Here was a man of my own heart. I would prefer it, I told Coccia, if science categorized species one beside the other. That would still allow an order, a system of sorting, without imposing any kind of a hierarchy. He immediately agreed. He reiterated his belief that the ordering system we have today is not scientific but rather influenced by cultural, historical, and religious values. For Coccia, the hard boundary between the plant and animal world does not exist. He believes plants can experience sensations and even reflect on them. And he is not the only one who thinks this, as you will see as you read on.

Accepting recent insights into the self-awareness of plants and doing away with a qualitative ordering system not only meets with resistance in conservative scientific circles but also leads to new emotional problems. A typical question that comes up is "Then what is there left for us to eat?" Meat eaters can't resist making snide remarks about vegans who, according to this way of thinking, are also inflicting pain with their exclusively plant-based diets. But that brings us to the topic of morals and, according to Coccia, thinking in moral terms is the greatest enemy of science. Basic research establishes facts without drawing any moral conclusions. Moral evaluation belongs as part of the political discussion that follows, as we know from decades of debate around raising animals.

Many people are clearly terrified about the consequences for their worldview, and so discoveries being made today about plants and what they can do are often dismissed as pure fantasy.

IN THE END, this is exactly what stands in the way of making meaningful connections with forests and trees. There is that little voice in your head that constantly cries "This isn't real!" In the beginning, bookstores often relegated my book *The Hidden Life of Trees* to the fiction section, even though it was based on facts. They did this because the way I write is considered unscientific. Some forest experts also found it too emotional, and therefore they did not take the book seriously. I am pleased to report that this rigid way of thinking is showing signs of loosening up, thanks, in part, to the efforts of many universities. There are increasing numbers of employees whose job it is to work with the media and translate scientific results into exactly this kind of language, one that is easy for non-scientists to understand so that the people who fund their research—in other words, the general public—can share in the results of that research. Progress is being made, but slowly, and there is still a tendency to discredit unwanted discoveries by banishing them to the mystical realm of esotericism.

At this point, it seems to me that it would make sense to explain what the term "esotericism" means. My dictionary defines it as "ideological movements, which have as their goal the self-awareness and self-actualization of people through the citation of, among others, the occult, anthroposophical and metaphysical teachings and practices." Other definitions state that esotericism is non-religious spirituality.

Let's stick with the dictionary definition and look at the individual components one by one. The occult has to do with supernatural phenomena and is sometimes used as a synonym for esotericism. If the terms are not exactly the same, there is at least a strong overlap between the two. That is a classic mistake when it comes to definitions. You cannot define a word using a synonym, and so we can cross the occult off the list of dictionary definitions. Do we fare any better with the word "anthroposophical"? Absolutely not, because it, too, contains a central supernatural component with respect to humankind and its development. That leaves us with metaphysics. And once again we are in the realm of things that cannot be proven—that is to say, the supernatural. I admit that metaphysics tends to concern itself with philosophical questions (for example, Does God exist? Who created the universe?), however, the word doesn't help much in defining esotericism. What we are left with is a nebulous idea of something spiritual and supernatural that in our day and age has negative connotations. When scientific research is described as "esoteric," the term is being used in a negative way, even though scientific research is the exact opposite of esotericism. After all, scientific researchers are not making assertions; they are setting out to prove things. This mode of criticism is sometimes used when revolutionary research threatens the stability of well-entrenched concepts.

People who use the term "esotericism" in this way are reducing the value of many things we currently find inconceivable. And there is so much we cannot yet grasp, especially when it comes to plants. Someone, however, who is very good at making inconceivable connections is František Baluška. I mentioned the professor from Bonn in a previous chapter, and in October 2018 I was finally going to meet him at his workplace. I excitedly imagined what his kind of plant research would look like: well-equipped laboratories with plants all over the place monitored by elaborate apparatuses finally giving up their secrets. That was something I absolutely had to see.

I PARKED MY car in front of the institute on a sunny afternoon. I took a musty-smelling elevator to the fourth floor. Then (according to the email I received from Baluška), I was to turn right when I stepped out of the elevator and take a flight of wooden stairs up to his office. The corridor straight in front of the elevator door led to neat, uniformly gray rooms of the kind you expect to see in universities. The wooden staircase to the right led to an out-of-the-way corner in the huge building complex. Up there, on a tiny landing, Baluška greeted me with his strong Slovakian accent.

He led me into the conference room and we sat side by side at a huge round table. I was eager to hear what he had to say. After all, I had cited his research in my book about trees and repeatedly mentioned his groundbreaking research when I attended events. His results sounded so fantastic that I sometimes wondered if I had interpreted them correctly when I translated them into everyday language for the general public. Baluška immediately put my fears to rest.

One of the first things we talked about was how plants feel pain. Fellow foresters roll their eyes when I talk about spruce feeling pain when they are attacked by bark beetles. "Of course a plant, trees can feel pain," the professor answered when I asked him about it. "Every life-form must be able to do that in order to react appropriately." He explained that there is evidence for this at the molecular level. Like animals, plants produce substances that suppress pain. He doesn't see why that would be necessary if there was no pain.

Baluška was ready with other quite different discoveries, as well. There's a vine that grows in South America that adapts to the form of the tree or bush it is climbing on. Its leaves look just like the leaves on the host plant. You might think this is chemically controlled. In that case, the vine might be detecting scent compounds from the bush and changing the shape of its leaves in a way that was genetically predetermined. Three different leaf shapes had been observed. Then a researcher came up with the idea of creating an artificial plant with plastic leaves and relocating our botanical chameleon to its new home. What happened next was amazing. The vine imitated the artificial leaves, just as it had imitated the leaves in nature. For Baluška this is clear proof that the vine can see. How else could it get information about a shape it had never encountered before? In this case, the usual suspects—chemical messages released by the host plant or electric signals between both plants—were absent. He went further. In his opinion, it is conceivable that all plants might be able to see.

Up until then, the only thing I knew was that trees can differentiate between light and dark. Sleep behavior has been researched in birches and oaks, and beeches can measure day length—all of this requires light receptors that transmit signals to the trees and spur the whole organism into action. This is far removed, though, from vision in the sense of being able to recognize shapes and colors. And now this: plants which register precisely that and change their behavior accordingly. I found that astonishing.

Baluška directed me to research being done on the cuticle or outer layer of leaves. On most plants, this layer is completely transparent, which makes no sense if all the leaves are doing is collecting light to manufacture sugar. In that case, these outer cells should be equipped with green chloroplasts, the organs used for photosynthesis—after all, this is where the most sunlight falls. Logically, less light is harvested in layers farther from the surface. And yet the cuticle is transparent, which seems wasteful. Not only that. In several plants, the cuticle is constructed in the shape of a lens, which means that it focuses light, making the cuticle functionally similar to the lens in our eye. It doesn't seem logical to me to focus light if photosynthesis is the only goal, because the cuticle could simply let the sun's rays through. Focusing light doesn't increase how much light falls on a leaf. The same amount of light is simply more concentrated or, more specifically, focused more intensely to the back of the cell.

LEAVES THAT FUNCTION like eyes? There's an idea that takes some getting used to, particularly as a tree regularly discards its "eyes" in the autumn when its leaves fall off. Does that make leaves disposable eyes? In a certain sense, yes. A working life of six months (under European climate conditions) is relatively long in comparison with some animals. Flies, for example, use their eyes for little more than a month simply because that's how long they live. And mayflies, which live for barely a day after metamorphosing from a larva into a flying insect, use their visual apparatus for less than twenty-four hours—and yet the eyes they have are real.

There's another thing with trees. The cells in the leaves, once they are formed, last for the whole growing season, which means they are relatively long lived. In contrast, our eyes are in a constant state of partial rejuvenation: the cells in the outer cornea, for instance, are completely replaced every seven days.[64]

You would think that plants experiencing pain and now the hypothesis that they might even be able to see would put the whole scientific community into a state of high excitement. The reaction, however, was muted. I had assumed that plant neurobiology was an up-and-coming scientific field. Baluška shook his head. "Germany was once very active in this area of research. Recently, though, there has been hardly any money for further studies." He was practically the only one still studying the topic in depth. And that means this branch of science could disappear and be forgotten for a second time. The first time it disappeared was back in Darwin's day.

Charles Darwin had studied plant roots and even back then he postulated that the tips might function like the brains of simple animals. Roots containing "tiny brains"? The hard boundary between animal and plant could have fallen in his time. Could have. The research was put on hold for a hundred years and then suffered another blow that it has not recovered from to this day.

The blow came in the form of a well-intentioned book by Peter Tompkins and Christopher Bird entitled *The Secret Life of Plants*, which was first published in 1973. It was based not only on proven facts but also on experiments that were not reproducible and definitely strayed into esotericism (there's that word again). No matter what you think of this book, it set investigations into stimuli and information processing in plants back for decades. But it would be unfair to place all the blame on this book. For one thing, research into these topics should not have been affected. After all, the book represented only one of many different opinions. For another, the profession seemed to have been waiting for an excuse to cut off this rogue branch of knowledge.

THERE WAS, HOWEVER, another, completely different, problem, as Baluška explained. All the research on nerves, the brain, and phenomena such as pain had originally been done on people. All the important biological terminology, therefore, was already taken. This meant it would not be scientifically correct to transfer the definitions to plants that exhibited very similar structures and processes. And so, neurobiology was reserved for animals, which is why a similar periodical for plant research is called *Plant Signaling & Behavior* and not "Plant Neuroscience." I immediately thought that philosophy and biology should be more closely connected, because Coccia's thoughts on the subject matched those of Baluška.

THE PATH TO greater harmony between all life-forms is long, as we see and hear every day. The words we use reflect our different approaches to plants and animals. We know what animal protection means: it encompasses everything that helps to protect animals' needs both legally and in practice. Factory farms don't belong under this definition, nor do medications administered to the animals because so many of them are packed into tiny stalls.

Things are quite different with plants. Crop protection doesn't mean that plants are being protected. What this term covers is the part of conventional agriculture dedicated to reducing losses caused by competing vegetation, insects, or fungi by every means possible, including toxic chemicals. Glyphosate is a magic bullet that exists solely to kill plants.

CUTTING DOWN TREES is another example. Here's a little thought experiment to show how the meaning of words has been twisted when it comes to forests. How would you react if butchers were described as being in the business of caring for animals? They would say they were removing excess animals from barns so the remaining pigs and cattle have more space. The remaining animals are now free to develop as members of their species should and their ranks will be constantly rejuvenated, both of which have positive effects on the health of the group as a whole. That probably sounds weird to you.

I think it shows butchers could learn something from foresters when it comes to public relations. Trees, after all, are almost as beloved as large mammals such as elephants, and most people want to treat both with great care and protect as many of them as possible. Accordingly, people find heartless behavior toward these beings outrageous. And this leads foresters to come up with harmless-sounding names for harsh interventions. Foresters describe "thinning" as caring for the trees, when what it really means is that up to 20 percent of the trees in a stand are cut down (that is to say, killed) and then processed. The foresters argue that the open space created by thinning benefits the remaining trees. But these trees don't need more space. What they need is a well-functioning social community.

In the foresters' defense, it has to be said that they raise the wood we all want to consume, such as the wood used to produce the paper for this book. And for that trees must die, which, in turn, hardly anyone wants. Foresters in Germany call this the "slaughterhouse paradox" (and in so doing, they really do compare themselves with butchers): many people like eating pork chops, but no one likes being confronted with the suffering and slaughter of pigs. What foresters have not yet understood, however, is that their practices make them part of the problem. What forest agencies need is not publicly palatable ways of describing what they are doing, but a different way of understanding nature. Only when they admit that what they do is exploit rather than protect nature can a real and open discussion on the topic of forests and wood products take place.

THERE IS NO way we can use the forest or nature in general without destroying something. The question is simply how much we want to ask of our ecosystems. This is a difficult question, and it has a lot to do with giving things up. The less wood we use, the more forest can be protected.

Along with my team from the Forest Academy, I work the forest and, yes, we also cut down trees. I know that what we are doing is for the benefit of people rather than the forest. And so, our guiding principle has to be to do as little harm as possible and to disturb natural processes only when we have to. What this looks like in practice is to stop planting trees and to log what grows there naturally. Where I live, this means harvesting deciduous trees such as oaks and beeches, mixed in with a selection of other species such as hornbeam and maple. Clear-cuts and applications of insecticide are out of the question, and horses win out over heavy machinery. Trees are, and should be, allowed to grow old on at least 10 percent of the forest area. And yet, even with these ground rules, I am still not a protector of the forest but a producer of wood.

THE MORE I think about the differences between conventional forestry and the original ecosystems of the forest, the more I conclude that the differences rest on a big misunderstanding. Conventional foresters believe that they are protecting ecosystems and through their stewardship are imitating or at most speeding up natural processes. However, the understanding of these ecosystems is grounded in a different philosophy about natural processes, in short in a different definition of evolution. This definition goes back to Darwin and his colleagues, who coined the phrase "survival of the fittest." However, that doesn't mean every life-form fights every other life-form and the strongest prevails. Rather, it's more about being able to thrive in an environment and reproduce successfully. That is a completely different interpretation of "survival of the fittest" and means, for example, that social communities can also be very successful in nature.

Trees and wolves—and especially our own species—prove how successful social communities can be. A more accurate rendering of the phrase would be "survival of the most well adapted" ("fittest" in the sense of being the best fit rather than the strongest), which means survival of the species that manage best in the environment in which they find themselves. If that were not the case, evolution would mean that ever stronger and therefore perhaps also more aggressive species would be the ones that survived. Moreover, if you read the phrase as the strong species surviving best, you would expect earlier species to have been underdeveloped, whereas in reality they were well adapted to the conditions that prevailed at the time. But because nature is always in flux, continents wander, and climate changes, the appearance and disappearance of species is not evolution in the sense of improvement but simply in the sense of adapting to new environmental conditions.

I, for one, used to interpret the phrase completely differently and thought that species were constantly improving until we finally got to us. And so, according to this outdated understanding, the logical conclusion was that humans stood at the pinnacle of creation. From a scientific point of view, however, this conclusion is incorrect. It's current meaning can only be explained from a cultural and religious perspective. And when we get to trees, we have got the wrong end of the stick completely, just like many foresters.

Foresters believe that trees not only of different species but also of the same species fight each other for light, water, and food. In managed forests, foresters get involved in what they think is the fight that plays out in undisturbed forests. You could say they see themselves as the referees. In Germany, I have often heard them say that the native forests could not survive without foresters. And yet, trees have existed for more than 300 million years, modern humans for 300,000, and the profession of forestry for just 300. For most of the time, trees have managed very well without human referees—in no small part because they have not been fighting.

And here I return to Coccia. He thinks it is a great shame that for the past one hundred years we've seen nature as a huge war zone in which everyone is fighting everyone else. But, according to Coccia, nature is not a war zone. On the contrary, it is characterized by solidarity. To that thought, I have nothing to add.

16

THE LANGUAGE OF THE FOREST

OUR DEEP RELATIONSHIP with the forest is echoed in language. The book you are holding in your hands reveals an early connection. I don't mean my book in particular but the word "book." If you trace its origin, you come to the Brothers Grimm of fairy tale fame. In a dictionary of the German language that they published in 1860, they mention that old German characters were scratched onto wooden boards. And because these boards often came from beech trees (*Buche* in German, which is pronounced *"boo-huh"* in English), the name for such writing tablets was transferred from the tree to the functional object—*Buch*, the book.

But the term could possibly have arisen much earlier, when runes were carved into wooden sticks made of beech. In German, where letters are *Buchstaben* (*Buch* = beech and *Stab* = stick), this gets us one level deeper, for with *Buchstaben*, it is much easier to see the origin of the word. Although neither this nor the assertion made by the Brothers Grimm has been proven with absolute certainty, I like the idea that every book has us looking back to the forest.

Whereas the word "book" is pronounced in almost the same way as the German word for "beech tree" (and in German differs from the word for the tree by only one letter), the origin of other words is more difficult to track down. Take the word "true." It, too, has to do with trees, specifically the oak. The wood of oak trees is hard and resistant to weathering, just as human relationships should be, figuratively speaking. The original word in Indo-European is *dru*, which means "oak." In English it turns up as "true," and in words such as "trunk," a wooden chest in which important things are kept safe.[65]

REFERENCES TO THE forest can be found in idioms, as well, even if some have now fallen out of fashion. "She's shaking like a leaf," for instance. Leaves, particularly the leaves of quaking aspen, tremble when the wind blows through them. Aspen leaves have stalks that allow them to twist in the lightest breeze. This might allow them to gather more light so they can produce more sugar. Whatever the reason, no other tree has this striking response to wind. But who today still encounters quaking aspen? A long time ago, the rustle of aspen leaves must have been so common that everyone had a good idea how much a person had to tremble to resemble the tree.

Old place names also reveal our deep forest roots—although it's probably more accurate to say that old place names reveal how we have uprooted the forest. In the dim and distant past, the residents of settlements chopped their way through forests to make space for buildings and agricultural fields. In the middle of the eighth century, Central Europe was still 90 percent forested and all the forest was primeval. There was no form of forestry at that time because it was not necessary. Population density was low and forests seemed practically endless. Areas for agriculture, in contrast, were in short supply, and it took a great deal of effort to wrest them from the clutches of nature. Not only were the trees in the way, so were their roots. Each and every one had to be dug out and dragged away by teams of oxen. Without this preparation, plows would have gotten stuck every few yards. It's little wonder our ancestors decided to use place names to memorialize their laborious clearing of the land.

Some of the place names even reflected the clearing method used. In the German-speaking Alps, if trees were simply felled and burned, leaving the roots, *schwenden* (which means "slash and burn") might be part of the name. This quicker method was not suitable if you wanted to plant crops, but it was fine if you were not going to plow the land and all you needed was a pasture for livestock. *Schwenden* (the past participle of which is *schwand*) appears either as a place name in its own right (as it does in Baden-Württemberg and Bavaria, both of which have towns called Schwenden) or as part of the name (as it does in settlements such as Herrenschwand or Untergschwandt). Other variations can be found in city names such as Bayreuth. *Reuth* is another term for *Rodung*, and Bayreuth means a clearing in Bavaria or a clearing made by Bavarian people. Then there is Stockum, where the name tips its hat to the tree stumps (*Stock*) that remain after the trees have been felled. In the British Isles, names ending in *-lea, -ley, -leigh,* or *-leah* indicate forest clearings, and *-thwaite* indicates a forest cleared for tilling, often with a dwelling on it.

THE INFLUENCE OF conservative science is clear in the latest terms used to describe nature. Emotions are out; long live technical descriptions. Thus, the workings of the wonderful network of life are dubbed "ecosystem services." That sounds less like paradise and more like a heading from a contractor's catalog. And that connects neatly with the discussion I had with Emanuele Coccia—all creatures are servants of humankind. They have services to offer and must accept their assigned place in the rankings. They earn our protection based on their contributions to our well-being.

Even if we are not aware of it, we find it almost impossible to avoid subconscious emotional responses to words. The journalist George Monbiot described this very well in an article he wrote. If Moses had promised the Israelites a land not where milk and honey but the secretions of mammals and the vomit of insects flowed, Monbiot asked, would they have followed him? He argues for a different way of speaking and a new terminology that touches our hearts so issues around environmental protection finally gain traction.[66]

A case in point are phrases that are constantly misused by lobbyists in current debates around protected areas. Thus, in Germany, forests that are being designated as national parks are "decommissioned," to use the official term. What does "decommissioned" conjure up in our heads? It brings to mind something we don't need anymore. A fleet of vehicles that has become obsolete, for example. Things that are decommissioned are things that we no longer use and, above all, they are things.

A forest, in contrast, is a living organism and, as such, cannot be decommissioned and certainly not by us. Intellectually, we grasp what is meant by this term: no more trees can be cut down. In reality, it is only the heavy tree-harvesting machines and chain saws that have been decommissioned, while people are expressly invited to enjoy an area that is now free to return to nature. Mammals, birds, and insects arrive in far greater numbers than when the area was a dreary working forest, and none of them are obsolete. In contrast to a fleet of vehicles, a national park after the forest has been decommissioned is far more active than it was before.

So what word should be used instead? A protected area? An area, then, that we need to protect? From whom? The answer is clear: from us. The term "protected area" reminds us that we are the ones that are being kept out (even though we are only talking about a certain profession). The term brings with it a subtle undertow of guilt that is not beneficial in the long term. Many environmental groups have realized, quite correctly, that alarmism and the continual cries of doom and gloom tend to exhaust us instead of helping us rethink the situation.

My recommendation is this: untouched forests should simply be called "forests." That is not necessarily an improvement, but there is a second part to my recommendation. It means that everything else must be relabeled "working forests" or, even better, "plantations." Other countries have no problem with this. Oil palms in Borneo, plantings of eucalyptus in Portugal and Brazil—it is clear that all of these are plantations. In Germany, however, the dreary, uniformly aged plantings of usually non-native trees are called forests. Local forest agencies avoid the term "plantation" like the plague. If it were used, however, the general public would have a much clearer idea of how little real nature we have left around us. So we don't notice, managed plantations are called "forests" because this word has many positive connotations.

This sleight of hand with language exists in the United States, as well. There, lands designated as "national forest" are not simply places where you can go out and enjoy nature. They are expected to contribute financially, as well. Many parts of them are not wild places free to develop in their own time, but are managed spaces where timber is harvested and money flows to federal coffers.

The word "forest" carries an aura of wildness with it. Other words are studiously avoided. Fellow foresters are outraged when I compare foresters with butchers in my presentations. But what is felling a tree if not slaughter? The only difference is that a tree is killed, not an animal. If we know from the latest research that beeches and oaks are also capable of feeling pain, then it makes sense to use the same terms we use for animals.

Yet the covering up, the de-brutalization, has also slipped into our daily speech. Isn't wood a piece of nature (and therefore something positive)? Doesn't it still live and breathe even after it has been processed? That sounds like a second chance, a sort of rebirth as a living-room suite or dining table. Processed wood is completely dead, of course. All it can do is absorb moisture and then dry out again. Earthenware pots, clay plaster, and bricks all do this, too. Just to be clear: wood products are in and of themselves beautiful, and they remind us every day of the ecosystems to which we belong, but going into raptures about wood products does nothing to advance current science and makes it more difficult to seriously argue against modern forestry.

What we need, therefore, are not new words but simply more honesty. And if people still want to go out into their nearest forest, the real forest, but find only mechanically exploited plantations, then perhaps they would advocate more strongly for more protected areas so that they could find nature on their doorstep. That is something we can all hope for.

17

DIVING DEEP INTO THE FOREST

When I was a child, I loathed going for walks in the woods. I trotted along after the adults hoping I'd find a stick I could chip away at to lessen the agony. The reason for the walk was usually to take guests to the local tavern, where at least there was lemonade. My idea of a good time in the woods was being let loose to build dens with my friends or light a fire (even though that was strictly forbidden) or dig for treasure. A walk felt like an outdoor prison sentence.

Today, of course, I feel differently, and for many people walking in the woods has once again become popular. Ultimately, it all comes down to packaging, and this comes in various forms to appeal to different markets. The alternatives include trekking, hiking, Nordic walking, and forest bathing. They are all different, but they share the same goal: getting people back out under the trees. Except for forest bathing, they are all various forms of sporting activities. Pounds must be shed and as many calories as possible burned to make an outing worthwhile.

Surprisingly, researchers have discovered that the speed at which you cover a given distance is not particularly important. If you cover 2.5 miles (4 kilometers) at a walk, you will burn about 240 calories. If, however, you jog the same distance, you will get to your destination twice as fast but will burn a mere 80 calories more.[67] To put it another way: walking is better exercise than it seems. The great advantage of this form of locomotion is that it's easier to coordinate your legs and feet when you're moving more slowly, which means you have time to look around and enjoy the forest. Walking, therefore, is much more relaxing than jogging.

There is another reason, however, that walking under trees is beneficial: the compounds beeches, oaks, and many other trees release into the air to communicate among themselves. These affect our circulatory system and our subconscious, and our blood pressure sinks—although not in every forest. Experiments at the end of the 1970s discovered that while native deciduous trees in Germany show positive effects, our blood pressure can rise in non-native spruce or pine plantations, where stressed conifers exchange chemical messages about insect attacks and lack of water—and we likely pick up on these even though we are not consciously aware of them.[68] Your subconscious translates this arboreal activity into changes in your body as well as in your mind. The forests we experience as beautiful are those that are pleasant and lower our blood pressure.

I ONCE USED the German television presenter Bettina Böttinger to test how forests affect us. First, we walked around

Cologne. With the television cameras running, I took an initial measurement of her blood pressure among the high-rises, food joints, and trams stops. I was a bit nervous because this would be a quasi-official record of whether the whole thing really worked. Her pulse and blood pressure were elevated, but that didn't mean much—after all, I didn't know what Böttinger's baseline readings were. Then we drove out into the Bergische Land, a deciduous wood of oaks, hornbeams, and beeches. There we took the measuring devices out for a second time, the camera team got in position, and I looked at the readings with anticipation. Bingo! They were measurably lower. The trees were clearly as relaxed as she was.

Naturally, a one-off demonstration like this doesn't replace scientific studies. But these have existed for some time now and they continue to produce results. They concern not only blood pressure but also our ability to fight off disease. A walk in the woods does more for your immune system than you might think: it benefits from the defensive strategies of the trees.

Way back in 1956, the Leningrad biologist Professor Boris Tokin demonstrated that conifers did a good job disinfecting their surroundings. He discovered that the air around stands of young pines was almost germ-free. The cause of this clean air was the trees themselves, which were giving off phytoncides, a kind of plant antibiotic.[69]

Why do conifers do this? The answer is that they are constantly being attacked by an enemy we cannot see, one that drifts in the breeze. Every 35 cubic feet (1 cubic meter) of air carries up to ten thousand fungal spores just waiting for their chance to land on a broken branch or damaged bark.[70] From there, the fungus grows into the tree and slowly eats it from the inside out. The wood rots and the tree dies. It's understandable that many conifers want to defend themselves at the earliest possible opportunity, preferring to take out the attackers before they even land. (Deciduous trees deal differently with fungus, as I learned on a trip to the ancient Polish forest of Białowieża, but more about that later.)

Conifers fight fungal spores before they reach their bark, and people who suffer from allergies benefit from the trees' preventative measures. But people who suffer from allergies are not the only beneficiaries. Unbeknownst to you, you breathe in the trees' defensive compounds, the phytoncides, with every breath you take, and they help protect your body, too. In your case, they trigger a reaction that reduces inflammation. In addition, phytoncides have been found to reduce the activity of cancer cells. Japanese researchers at the Nippon Medical School discovered this when they sent test subjects out into the forest or the city. Cancer-killing cells and anti-cancer proteins increased in those people who visited the forest but not in those who visited the city, and the elevated concentrations of both could be detected in the participants' blood up to seven days after the forest walk.[71]

Researchers in Korea ran similar experiments to test this phenomenon. They had a group of older women walk in the woods and another group walk in the city. The results were astounding. The blood pressure, lung capacity, and elasticity of the arteries of the forest group improved, but there were no measurable changes in the city group.[72]

THE TERM "CITY" is perhaps a bit vague when it comes to health. Apart from noise and air pollution, there are other biological characteristics that differ considerably from suburb to suburb. This brings us back once again to trees. Various studies ascribe clear health benefits even to street trees. In a large-scale study, scientists at the University of Chicago discovered that a single tree by the front door improves health and well-being. They gathered data from about thirty thousand residents of Toronto—and from 530,000 trees the city had already mapped. The result: ten more trees in a residential neighborhood improved the health of the residents as much as an increase in income of $10,000 a year (including the improved medical care that comes with such an increase). And we are not only talking about mental health. The likelihood of heart and circulatory diseases, the leading cause of death in North America these days, dropped measurably. Eleven more trees in the neighborhood was an improvement in cardio-metabolic health equivalent to an additional $20,000 a year or, measured another way, it reduced a person's biological age by 1.4 years.[73]

TREES ARE GOOD; forests are even better. The German physician and talk show host Eckart von Hirschhausen told me that in Japan today, a doctor can write a prescription for a walk in the woods—a sick note, as it were, that gives you permission to spend time in the forest. And the latest trend from Japan is making its way over here: forest bathing.

I must admit that I was somewhat skeptical when I first heard about it. You can't bathe in a forest. How is that supposed to work? Okay, so it has to do with relaxing however you want, but hasn't that always been possible? Is forest bathing simply a case of repackaging an age-old practice? Despite all the new books on the subject, it wasn't clear to me exactly how it differed from traditional outdoor activities. And why was it suddenly so popular?

It could be because attitudes have been changing in recent years as people are finding their way back to nature. It reminds me of German children in the 1970s and '80s collecting corks from wine bottles and the foil tops of yogurt cartons to save bark from cork trees and energy in aluminum mining and production. There were similar drives in Britain, where the children's television program *Blue Peter* had children collect the aluminum tops from glass milk bottles. Luckily, aluminum bicycles and alloy wheels hadn't caught on yet or we would have felt like Don Quixote tilting at windmills.

After the fall of the Berlin Wall in 1991 and the reconciliation between East and West Germany, the focus switched from conserving resources to economic growth and worries about the upsurge in terrorism. Nature became less important to young people. At least that was always my impression from the tours I led in my forest. But in recent years, the desire for an unspoiled environment has returned, as you can see from the growing number of citizen initiatives to protect forests everywhere. I'll come back to these later.

With this longing for natural spaces, forest bathing has spilled out of Asia. Called *shinrin-yoku* in Japanese, the whole thing sounds like ancient wisdom. However, it isn't at all. Quite the opposite, in fact. Japanese forest agencies came up with the idea and the name in 1982 as a way to make people more aware of the health benefits of the country's forests.

Japanese foresters were not the ones who invented restorative walks in the woods. The Bavarian priest Sebastian Kneipp discovered the health benefits of nature back in the nineteenth century. He had tuberculosis as a child and healed himself by taking cold baths in the Danube. After he entered the priesthood, he continued to study alternative health practices. In addition to cold water, which he used in many different ways, he employed a wide variety of herbs. Many doctors and pharmacists sued him because what he was doing did not align with traditional medical knowledge and, what was even worse, he was offering his services for free. Kneipp was, however, acquitted and continued to share his methods with his patients. Naturopathy, as it is now known, is practiced to this day.

Forest bathing started in a similar way. People already knew that going out for a walk in the woods was a relaxing and healthy thing to do. It's just that no one knew exactly why. The explanation came, as I have just described, with the scientific discoveries of chemical messages and other compounds released by trees.

It's clear, then, that we react physically to forests. However, it was still not clear to me what "forest bathing" really meant, either as a new outdoor activity or as a therapeutic practice. It's certainly helpful to know more about one of the pioneers, Dr. Qing Li of the Nippon Medical School in Tokyo, Japan. The professor teaches and carries out his research at this small private university of under a thousand students. The school is well regarded in Japan and groundbreaking research is done there. Li is far from being a romantic dreamer. In 2018, he published a three-hundred-page book entitled *Shinrin-Yoku: The Art and Science of Forest Bathing.* In it he describes not only his research, but, most importantly, how forest bathing works.[74] And it's really simple. You choose a forest you like (it could even be in a city park) and you go there to relax. Then you gather all your senses and dive in to the smells, sounds, and sensations. According to Li, all you need to do is accept the forest's invitation to slow down. Mother Nature takes care of the rest.

If this is all there is to it, you don't need a book. A small brochure would suffice. As you read on, Li suggests a number of activities, including listening to the birds or becoming aware of different shades of green. He also gives tips for what to do immediately after forest bathing. He suggests a tea ceremony, for example, or burning incense to bring the scent of the forest into your home.

If you're left with the impression that I consider the book unnecessary, I should explain myself better. These days, we have completely forgotten how to engage with the forest. We no longer take the time to do nothing but wander under the trees or lie down for a couple of hours on the soft forest floor. If we did that, we'd be regarded as slightly eccentric. The picture changes, however, when everyone is doing it as part of an accepted program with recognized health benefits. And that is what forest bathing is for me: it gives me permission to go out and relax under trees.

We experienced a similar sea change in behavior when Nordic walking elevated a walk in the woods to an official fitness program. Apart from walking shoes, all you needed were special poles. The idea came from a physical education and sports student in Finland who met a man who made ski poles. The ski-pole manufacturer was complaining about the lack of a market for his products in the summer and was inspired by the student's training program. The ski poles were quickly modified and all was ready for a mass movement. Sales were brisk, even in the warm months, and since then the poles have been gouging millions of holes along the sides of forest trails.

I want to be clear here. This kind of physical training makes a lot of sense because it includes the upper body. It also motivates a lot of people to train outdoors instead of at the gym. And forest bathing leads to similar enjoyment outside the four walls of your home. Courses led by guides confer an additional benefit. If you're alone in the woods it doesn't take long before you slip back into your normal rhythm and that might put an abrupt end to your experiment in slowing down. If you're taking a course, however, you usually stick with it until the end (even if it's only because you've paid for it), and helping people stay with the program is one of the reasons I've decided to offer a course in forest bathing at my Forest Academy.

PERHAPS I SHOULD have done more forest bathing earlier in my life. I could tell you a thing or two about how hard it can be to let go. In 2008, I got burned out. Specifically, I fell into a depression brought on by exhaustion. I had been feeling anxious and tense for weeks, because I was completely overwhelmed by my forest job. It wasn't anything my employer was asking of me. I was the one who wanted to do everything and then more to protect the forest. And so, I kept adding other projects in addition to my full-time job as the forest manager without thinking of the consequences.

In order to spare the ancient beech woods, I established one of the first forest burial grounds in Germany, where you can choose a tree under which you will later be laid to rest. Apart from getting a maintenance-free grave, you will also be ensuring that the old deciduous forest will not be cut down for at least another ninety-nine years, which is the length of the lease on the burial plot. In addition, I launched an ancient forest program where additional areas of ancient beech forest could be leased by the square foot with a click of the mouse. My goal was to save as many of the last ancient deciduous forests as I could in the place where I lived. I also organized seminars, tried to raise hunters' awareness of the environment so they at least stopped shooting foxes, gave presentations to conservation groups, and worked with scientists. When both of my coworkers dropped out due to illness and I had to take on their workloads as well, my body said, "No!"

It made its announcement in the form of panic attacks during a live broadcast on a local radio station. Every ten minutes, a crippling wave washed over me and every time, I thought my heart would stop. Somehow, I managed to keep it all together for the duration of the broadcast. From the outside everything looked fine; on the inside I was dying a thousand deaths. Later I went into therapy for a few years to learn to pay attention to my needs and to dial back my plans to save the forest, at least a little. I can hear you saying, "But aren't you on the go all the time right now?" I admit it. You're right. I've got a lot on my plate once again, but this time I'm much better organized. I've reduced the area of the forest I'm responsible for, and two colleagues have assumed most of the workload. My son, Tobias, now runs the Forest Academy, and my agent handles the requests I get from around the world. Miriam is in charge of my appointments, which means I once again have two days off a week. I no longer ignore my body when a slight disruption in the rhythm of my heartbeat sounds the alarm, and if I'm in any doubt I say no a lot more, something that I still find hard to do.

WHAT HAS ALL of that to do with forest bathing? Miriam and I live in an old lodge in the middle of the forest, and I also spend a lot of time outside under the trees. If there are such clear connections between health and how trees affect our bodies, how did things get in such a bad way with me?

First, of course, is the question of how much worse things might have been for me without the forest. Then there is the fact that the beneficial effect of the trees cannot help beyond a certain level of self-destructive behavior. I have learned some things since my burnout, but I still find it difficult simply to go out into the forest to relax. Once, though, it did work.

My children gave me a family walk through my forest for my birthday. That might sound a little odd, but they knew that relaxing time together as a family out in nature was the most precious gift they could possibly give me. We dawdled along a narrow path, stopped at every flower with a butterfly on it, and snacked on dark-red cherries hanging from a tree by the side of the trail. After a short walk through a stand of deciduous trees, my children spread out a blanket and served an ample picnic. We lay there under the trees for one, maybe two hours—I can't rightly say any longer. We chatted and relaxed and forgot time. And that is forest bathing. For me it was the most beautiful day in the forest that I can remember. And that's saying something, because over the course of my life I have spent thousands of days in the forest. So in case you, too, don't find it easy to simply lie down on the leaves, even though you'd really like to, I can recommend this: go forest bathing with a guide.

I WOULD ALSO like to specifically recommend Dr. Li's book. It makes you want to step into the forest by first inviting you on a forest walk from the comfort of your armchair. And what would happen if all the people who read all the books about forest bathing decided to step out into nature? Is that something the forest could accommodate? Because I like to encourage everyone to spend more time in the forest and even to wander off the path, this question comes up often. Of course, too many people put a strain on nature, but if we compare the disruption caused by visitors with the negative effects of modern forestry, it is vanishingly small.

SINCE 2019, THE Ludwig Maximilian University of Munich has been educating forest health trainers and therapists and researching the science behind forest therapy, which means forest bathing has received a blessing from academia.[75] It shouldn't be long now before doctors in the West are also allowed to prescribe walks in the woods. And I hope that happens, not only for people's sake but also for the trees' sake, because then people will learn to value ancient forests. After all, who wants to go for a stroll through a dreary plantation if they have the opportunity to visit an intact, fully functioning forest?

18

FIRST AID FROM NATURE'S MEDICINE CABINET

If we're investigating the indirect effects of trees, we should also look at the direct health benefits these giants offer. People are skeptical, as I've seen time and again on my forest tours, but as long as you have a guidebook and can identify a few common trees, this is an area where you can experiment. You can, for example, eat the leaves of beech, oak, and many other deciduous trees with no worries at all. They're even good for you. But when I try to get my guests excited about trying them, they just look at me. You mean you can just bite into a leaf? Really? Yes, you can, and in spring, at least, right after the trees have leafed out, the delicate green leaves taste delicious and slightly bitter. You can concoct beautiful salads using only beech and oak leaves.

You don't even need to bring any headache medication into the woods, because willows offer you something similar.

Their bark contains salicin. Willow is *Salix* in Latin, so salicin is named after the tree. Depending on the species, willow bark contains up to 10 percent of this substance, which your body processes into salicylic acid after you have ingested it. The well-known synthetic medicines whose active ingredient is acetylsalicylic acid are more potent, but they also have more side effects, such as acting as a blood thinner. If you have a headache or a fever and you don't want to experience these side effects, you can reach for a cup of willow-bark tea instead. People have been doing that for thousands of years, as we know from clay tablets dating back to 700 BCE. Synthetic salicylic acid dates back to research done around 1830, when scientists figured out the secrets of willow-bark tea. The modern white tablets are nothing more than a chemical reconstruction of compounds present in our native trees.

Of course, it would be a shame if you were to go out into the forest right now and peel the bark off a willow. That would be like peeling the skin off a living animal. But if you cut off a couple of twigs and take the bark off them when you get home, you will minimize the damage to the tree. Germany's native white willow, which grows along river banks, is a particularly good choice. In the forests in the Mittelgebirge, the willow you come across most often is the pussy willow, which often grows at woodland edges or in clear-cuts. The beeches and oaks don't allow these little trees—they grow barely 50 feet (15 meters) tall—to grow in their shade, and the pussy willows find their niche here instead. It's true that they contain less salicin than white willows, but why don't you give them a try? And if you don't want to cut anything off a tree, meadowsweet can often be found along streams and in damp areas. This perennial herb, with its white flowers and swampy-sweet smell, contains similar compounds to willow bark. If you gather the flowers in June and July and make a tea out of them, you should also be able to feel their effects. Meadowsweet and willows are also common in North America.

THE FOREST IS helpful not just when you have a headache. How about something to treat insect bites and other swellings? For this you need a maple or, to be precise, one of its leaves. Lightly crush the leaf and lay it on the insect bite, and it will help keep the swelling down. This also works on your feet if they've swollen after a long hike.

Oaks, on the other hand, are most useful internally. For instance, if you have a sore throat. You need some bark, which you brew into a tea, and then sip. Now, as with the willow, I don't want to encourage you to peel the bark off oaks, because that would severely damage the tree. If trees have been cut in a wood near you recently, you might pry some bark from a downed trunk. A trip to the drug store or health-food store is even easier. There you will find oak bark tea already dried and packaged.

In spring, you can also make a tea from fresh spruce tips. They contain a lot of beneficial acids and vitamin C. At this time of year, the drink is reminiscent of lemon tea. Later, however, the bitter compounds increase and spoil the taste. And why wouldn't you just want to drink lemons? As with all the other uses I've mentioned so far, it is all about strengthening your bonds with nature.

Not that I want to suggest that you should live as though you were still in the Stone Age. But incorporating little things like this in your everyday life helps you understand the forest better and feel more connected to it. Moreover, the ingredients in the trees are neither sprayed nor processed in any way. And it's fun to gather forest products and easy to spark children's enthusiasm for this activity. You can give them a special treat, for instance, by making chewing gum. All you need is a clear, thick, stone-hard drop of resin from a spruce. You can identify spruce by its reddish-brown bark and long, hanging cones. Don't worry, the resin from pines, Douglas-firs, firs, or larches is not poisonous, chewing gum just works particularly well with spruce resin.

Put a solidified drop of spruce resin in your mouth to warm it up to body temperature. Test it carefully every once in a while to see if it's soft yet. Don't bite down on it too hard. If you do, the resin will shatter into a thousand pieces and immediately release all its bitter compounds. So, take your time and gradually begin to chew. Spit the bitter compounds out when you taste them (now you know why you should undertake this activity only when you're out in the woods) until finally you have a chewable, rosy-colored mass—it's ready. The tastiness of this gum is debatable, as it never loses its resinous undertone, so I suggest you consider this activity primarily as a star turn during a walk if you're the trip leader and want to offer your family or group of friends something different.

Trees, incidentally, even offer something for the kitchen. Douglas-fir needles taste tart and bitter, a bit like candied orange peel, and are great for flavoring a wide variety of dishes.

THE ANIMAL WORLD has a few medicines ready for us, too. Insects such as bees even have an antibiotic on offer, waxy propolis, which is tree resin they've gathered from branches and buds and enriched with their saliva. They use propolis to disinfect well-traveled areas and to wrap foreign bodies (as large as a dead mouse!) in a sterile covering. They also use it to plug holes in their living quarters. Some beekeepers gather this natural putty from their hives and sell it, dissolved as a tincture, to the pharmaceutical industry as an alternative medicine.

On the subject of bees, if you are stung by them or by any of their wild sisters—yellowjackets, for instance—there's a herb you can use to ease the pain. It's called plantain. Actually, there are two different kinds of plantain: narrow-leaf plantain and broadleaf plantain. Plantain grows not only in fields but also along roadsides. This is most convenient, because it means that even if you're not out wandering through nature somewhere remote, in most places you'll easily find the help you need. If you crush a leaf until it's mushy and place it on the sting, it will reduce the pain and disinfect the site at the same time.

MEDICINE FROM THE forest is not a new discovery, of course. We all know that healing herbs have been used since the Middle Ages. The issue here is whether forest medicine is an example of how we are rediscovering an ancient bond with nature that existed long before the appearance of modern humans.

It might be helpful to take a look at the animal kingdom, especially our closest relatives. Chimpanzees, for example, have been seen eating the pith of bitter leaves as a purgative to rid themselves of gut parasites. But how did the researchers know the apes were eating the plants for their medicinal properties rather than for their calories? That was easy, because the leaves are toxic, even for chimpanzees. The animals seemed to know exactly how much they could eat safely, and after a purge they didn't eat these plants until the parasite loads in their guts increased once again. The animals obviously had a pretty good idea of what they were doing.[76]

You can picture apes self-medicating from nature's drugstore, but what about animals that are not as closely related to us? Take, for example, birds. To rid themselves of parasites, they not only use plants but also the services of other animals. Ants become their unwitting assistants when it's time to remove mites and other pests. To enlist the ants' help, the birds crouch down over one of the mounds made by these social insects and fluff out their feathers. The ants defend themselves against what they assume is an animal attacking their home by biting anything they cannot identify and, what's most important for the bird, spraying caustic acid. Between these two defensive measures, the ants kill large numbers of the parasites tucked away in the bird's feathers.

Eons ago, did our ancestors do something similar? We know that there are particularly hardy people today who lie naked on ant hills so the insects can bite them. It is meant to help if you suffer from rheumatism. Apart from the fact that this is forbidden in Germany because it is not an environmentally friendly thing to do, there is no evidence that it works.

THE CINNABAR MOTH practices a particularly unusual form of plant-based self-help. Its caterpillars like to eat an especially toxic plant—tansy ragwort. Tansy ragwort protects itself using what are called pyrrolizidine alkaloids. If a horse, sheep, cow, or goat eats the plant, it can cause death or at least severe liver damage. Every additional bite makes the situation worse until one time, perhaps years later, the animal takes a bite too many and dies. Tansy ragwort is equally dangerous for people, so it's not good news that its leaves resemble arugula closely enough that it is sometimes mistaken for it. In 2009, this led to a collapse in the arugula market in Germany after a shopper found a leaf from this toxic plant in a package of arugula salad.[77]

Cinnabar moth caterpillars, however, exploit ragwort's extreme toxicity. Although they eat other plants, they are irresistibly drawn to the toxic substances in ragwort. The alkaloids do not harm them. Instead, they spread through the larvae's tissues, making the caterpillars themselves deadly poisonous. That is their defense against anything that wants to eat them. And so that would-be attackers are aware of the danger, they adorn themselves with the same warning colors used by yellowjackets: alternating rings of black and yellow.

Although the cinnabar moth caterpillar is guided by instinct, the house sparrow provides good evidence that it purposefully uses substances in its surroundings as medicine. Dr. Monserrat Suárez-Rodríguez and her team from the National Autonomous University of Mexico were researching sparrow nests. They discovered that many of the birds incorporated cellulose from cigarette butts in their nests. There was an appreciable amount of nicotine in this material, which helped to significantly reduce the number of mites in the nest.[78] Because in this case the sparrows were not using plants—that is to say, natural medicine—this points to purposeful behavior.

USING NATURE'S MEDICINE cabinet is not a human invention but something that connects us to our fellow creatures. If we're rediscovering this natural medicine cabinet today, this is not a fad promoted by the environmental movement but simply a return to our roots.

While we are on the subject of roots, what do we do when the tables are turned and it is a tree that gets sick? Can we help? Are we even able to recognize that it needs help? This question is controversial to this day. And that makes it all the more interesting to take a close look at the arguments on either side.

19

WHEN A TREE NEEDS A DOCTOR

OUR LOVE OF nature often means we want to get involved and help when one of our fellow creatures falls ill. We especially like looking after species that are closely related to us (mammals) or life-forms that are particularly impressive, like trees. And because there are lots of trees around the places where we live, we single them out for our care and attention.

When old trees in the city begin to rot, we usually go on high alert. What is at stake here is not only the survival of a large plant but, more importantly, the safety of residents. A falling tree weighing hundreds of tons can do a massive amount of damage. Arborists arrive to assess whether the giant can be saved or whether it needs to be removed. In years past, arborists obviously spent too much time watching dentists: rotting trees were treated like rotting teeth. The rot was scraped out, and then the area was drilled and, finally, filled—not with a metal amalgam but with concrete. That sounds fine, doesn't it? After all, a concrete filling should provide a tree with the stability it needs. But a tree trunk is not a rigid structure. Wood is a combination of fibers and glue, and it flexes like fiberglass. A concrete core makes it impossible for a trunk to flex. It would be like inserting a steel rod in a human spine—you'd no longer be able to move around freely.

For the tree, restricted movement means the topmost branches are more likely to break off in a storm. Also, fungi spread much more easily under the concrete filler, usually because when the trunk was hollowed out, the inner partition layer the tree grew to protect its healthy wood was breached. That would be a bit like someone scratching the scab off a wound. Moreover, in rainy weather, the concrete gets soaked and slowly releases the moisture it has absorbed into the interior of the tree. This creates the conditions fungi like best, because it means they can grow undisturbed—directly into the healthy wood. From the outside everything looked great, but in reality, the decay and with it the danger posed by the tree were drastically increased.

TODAY THIS IS usually no longer done. Instead, the tree is put under strict observation. It is inspected regularly to find out how much healthy wood remains and if the tree is sufficiently stable. If it doesn't pass the test, its crown is carefully shortened to reduce the load the trunk has to support. This allows the tree to stay standing for a few more years. However, shortening branches always comes with its own set of side effects, as you will soon see.

Have you ever seen brutally mutilated street trees? They look as though sadistic tree cutters have had their way with the poor defenseless things. But the real reason they look this way is usually not nearly as interesting. The authorities were saving money. Crown reduction should be done by well-educated and, no less important, tree-loving people. They should be aware that they are cutting into sentient beings. Even if we can't imagine exactly what an injury feels like for a tree, it is a kind of pain, as Professor František Baluška has made quite clear. And if it is necessary to subject a large being to an operation like this, you should at least first consider very carefully how you are going to go about it and then keep the damage to a minimum. Unfortunately, this is the opposite of what you see in cities—and this is where I have to talk about money.

Tree experts are usually more expensive than municipal workers. In the autumn, after the leaves have fallen, there is often a lot of downtime when there's not a lot of work to be done. And because workers still have to be paid, it seems the perfect opportunity to use municipal staff to prune trees. The workers then grab their chain saws and do a particularly thorough job. If they're going to reduce the size of tree crowns, they might as well reduce them by a lot. Logic dictates that if they cut the trees back hard, they buy themselves time. It will take the trees many years to grow back to their old height, and until then, they won't have anything more to do. Unfortunately, that's not true. When the workers make their cuts, they unleash a cascade of consequences.

For starters, a tree subjected to this kind of treatment generally goes into shock, which is hardly surprising as major limbs have been amputated. A roughly handled crown reduction is comparable to having your legs amputated. Transportation channels for liquid inside the tree are quickly redirected, and as quickly as it can (which is usually not very quickly for a tree), the tree tries to seal off the wounds to protect itself from pathogens. However, that never works, at least not for wounds larger than 1 inch (3 centimeters). Fungal spores, which are found in every cubic meter of air, land on the cut surfaces within minutes and begin to germinate. Over the next few years, they will eat into the branch stubs and, bit by bit, compromise the stability of the whole trunk. At the same time, the tree now becomes extremely hungry. When large, living limbs are removed, the tree loses all the leaves growing on them, as well.

You might think that this loss is not such a bad thing, because the tree is also missing part of its body, which it no longer has to feed. But we are aware only of what we see. Underground, the tree has a root system tailored to its size that needs an enormous amount of energy. After the crown reduction, the root system can no longer be adequately supplied with food. As a consequence, many of the roots die off.

Crown reductions made to help trees withstand storms therefore often have the opposite effect. As the roots die, the tree becomes less stable. A new danger then looms. To survive, the tree grows bushy clumps of new shoots with unusually large leaves. It's hungry and leaves are the only way it can produce the lifesaving sugar it needs. These shoots later develop into branches. And because fungi are rotting the wound from which they are growing, this woody bouquet will eventually fall apart and create exactly the dangerous situation the crown reduction was supposed to avoid.

Now what? There's usually no way to save trees by cutting into them. But if you need to reduce the danger, it is possible to undertake a careful crown reduction by removing smaller branches far from the trunk. In most other cases, the only alternative is to remove the tree completely. Does that sound harsh? I think so, too. The solution needs to happen much earlier. Town planners and homeowners should think carefully about where they plant trees. Most importantly, if they have no idea how large the tree will be, they should call in an expert.

LET ME COME back once again to the dangers trees pose to people. In cities, it is mostly cars and houses—and their occupants—that are in the way of falling trees. In the country, people are out and about on roads, tracks, and hiking trails. Unfortunately, tragic fatal accidents happen. And although these incidents are extremely rare, they lead to complete overreactions classified as "the duty to implement safety precautions." What this means is that if you own one or more trees, you are liable for any damage they cause. It's much like owning a dog—it's a good idea to have liability insurance. The same principle applies to trees on your property.

Nervousness about potential property damage or, in the worst-case scenario, loss of life often leads people to err on the safe side and remove trees, especially in Germany where you are guilty of a criminal offense if a tree on your property strikes and kills someone. But, does the risk warrant the extreme reactions some people have? I'd advocate for saving as many trees as we can. The starting position is this: Trees promote good health, as I explained earlier in the book. If, as also mentioned earlier, the addition of eleven trees per residential neighborhood improves cardio-metabolic health equivalent to being 1.4 years younger, what's the situation with the life-shortening effects of unhealthy trees?

I looked through a lot of statistics, but I was unable to find anything useful. The problem is that many different causes get mixed in here. For instance, every year many people who use the roads are killed by trees. But when you look more closely, you discover that these accidents happen when a car leaves the road and tragically drives into a tree. Then there are natural catastrophes like storms, which regularly uproot trees that land on passersby. But there is not much difference between those accidents and those caused by falling roofing material. It's better not to step outside when a storm is raging if you can avoid it. Under such extreme conditions, I don't think it's fair to blame unhealthy trees. The complete number of accidents caused by branches that seem to fall from nowhere or whole trees that come down for no apparent reason should be relegated to a distant second place.

I'd like to put all this in the context of the population of Germany. Let's assume twenty deaths a year—which is on the high side according to news reports about such events—and an average age of forty for the people who are injured or killed. I'll calculate this as a percentage of the total population. Assuming a life expectancy of eighty, dangerous trees would reduce this by 0.00001 percent. The presence of trees in the city, on the other hand, raises this life expectancy by 1.8 percent. That is 180,000 times more. Even if slightly fewer trees were "cleaned up," which would reduce the gap between the numbers, there is still a huge difference between them. This calls into question the radical removal of supposedly dangerous trees. I am emphasizing this because traffic safety measures recently, here in Germany at least, are bordering on the fanatical.

EXPERTS ARE VALUABLE resources when it comes to evaluating the health of trees. If their services were used more often, more trees could be saved. Arborists who specialize in fungi that rot wood will come and offer their expert opinions on suspicious-looking fungi and either sound the all-clear or, if necessary, arrange for the tree to be removed. In addition, they evaluate the stability of trees whose roots have been damaged—by construction activity, for example. I myself have regularly used their services, and not only in the forest I manage. There is a pine tree in my garden that leaned over at a 45-degree angle after a storm decades ago. The tree is about 140 years old and correspondingly tall and heavy. Just how it can remain upright for even a day leaning over as it does is a puzzle to me—and to my neighbor.

To get clarification on this, I had a specialist who had been most helpful in my forest come over. He knows his subject so well that he has been able to save many trees. That is, by the way, a sign of a real professional. Less skilled appraisers call for all suspicious trees to be cut down to be on the safe side, just like the foresters do. This specialist, however, told me that everything was okay. The pine was so well rooted and had shored itself up so well that it posed no danger and could remain standing, which I was very pleased to hear. After all, the land around my forest lodge is a piece of the forest and should remain so.

A COMPLETELY DIFFERENT question came up for me when I visited the primeval Białowieża Forest that straddles the border between Poland and Belarus. I walked through the forest with Piotr Tyszko-Chmielowiec, who is both a scientist and a friend. There were dead trees lying all over the ground and enormous oaks and lindens whose 3-foot- (1-meter-) thick trunks were completely rotted on the inside and hollow like stovepipes. Up until that point, I had thought that the process of decay was definitely detrimental to living trees. That may well be true in many cases, but with these stricken giants, Piotr showed me that fungal attacks could happen for a completely different reason. In his opinion, old trees take a chance. They invite the fungi in and offer them their wood as food. That sounds a bit like suicide in slow motion, doesn't it? Not quite, if you follow Piotr's line of reasoning.

The pivotal factor behind inviting these parasites in is the tree's inability to move from the spot where it is growing. Over centuries, a tree's roots suck all the available nutrients, primarily minerals and nitrogen compounds, from the area around its base. At some point, this should mean there is nothing left and that would signal the end of life for the tree. When you think about it, after five centuries, a mature ancient forest locks up about 33 tons (30 metric tons) of biomass. It is locked up in the sense that the tree's living tissue and the decommissioned wood in the inner growth rings no longer contribute to the cycle of life, where life-forms are broken down by others to release the nutrients they contain. The tree continues to deplete the ground of nutrients and, eventually, all nutrients within the tree's reach are used up.

The trick to living another couple of decades or even centuries is to compost yourself. Fungi that enter via a wound in the tree convert the wood into a sort of humus as they eat their way through the tree, creating debris that is soft, crumbly, and moist. Now the tree can grow inner roots into this "soil" and reabsorb nutrients it stored in earlier years in its growth rings.

The first picture that popped into my mind was one of self-mutilation, but perhaps a comparison with rumination is more apt. Like a cow that brings up its stomach contents to chew them once again, the tree breaks down the inner part of its trunk and absorbs it for a second time. Unlike the cow, the tree is digesting material that once belonged to its skeleton. And that seems to be the crux of the matter. If a tree breaks down its supporting structure, won't it lose its balance and fall over? This is indeed the key question, and the answer depends on what part of the wood and how much of the wood the fungi attack. The innermost part of the tree—the oldest growth rings, which date back to when the tree was young—contribute very little to its stability. You can see that by looking at any steel pipe (the frame of a bicycle, for instance), which is hollow inside and yet retains its full load-bearing capacity. As long as a trunk is no more than two-thirds rotten, stability is not usually a problem.[79]

LET'S CONSIDER THE idea of nutrient recycling from a different angle. In October 2018, I visited Robert Moor, an author in British Columbia. We talked about the universal applicability of social systems. Is the social life of trees comparable to life in human communities? Might there be a common principle that underlies both? At first, I rejected this idea, because trees are more prone to sharing. Within a species, trees in ancient forests share sugar solution through their roots and warn each other of danger by releasing scent messages and communicating through their root systems. In short, there is no tree accumulating riches at the top. In human social systems, the same idea exists in principle. Through taxation, the rich contribute to the general pot, from which payments are distributed to poorer people. This means there is a certain amount of equalization, although it doesn't rise above a relatively low base level.

Mature trees share their strengths via the processes I mentioned, virtually eliminating differences in individual contributions. In human societies, however, there are huge disparities. Here, an individual like Bill Gates can pile up riches in amounts sufficient to permanently provide for a small country's total population. The forest is quite different in this respect.

Or is it? While I was chatting with Moor, I remembered my conversation with Piotr. Doesn't an enormous tree accumulate vast amounts of food? It does some sharing with its neighbors, but it keeps a huge storehouse of provisions gathered over centuries locked up inside its trunk and it has basically swept the ground at its feet clean of all available minerals. As it rots (whether willingly or not), it releases the nutrients once again and the humus that is created by this process becomes available not only to the tree but also to its neighbors.

The humus offered by the Bill & Melinda Gates Foundation comes in the form of a big bank account. Gates seems to be the rule rather than the exception among the super-rich. Whether they are actors, business owners, or sports celebrities, at some point, many rich people are no longer interested in having the most money. They do want to retain control of who profits from their fortunes and how, but they want to be rid of large parts of it—or they will be plagued by their social conscience.

Isn't that, in principle, the same as the distribution system in the forest? Both tree and human communities have a vested interest in stability, and inequality threatens this. What use is it to a tree to have a huge amount of food if its environment is suffering? If the forest weakens, then even a strong tree won't grow to be particularly old. Who will help generate the refreshing cool summer climate? Who will come to its aid if it falls ill? Every billionaire can ask similar questions about the social system in the country where he or she lives.

SO WHEN WE look at trees we think are sick, that are composting their inner wood, we should think twice before interfering. On the one hand, our own health depends on having trees around us. On the other, it is possibly further proof that nature is all about community not competition.

20

EVERYTHING UNDER CONTROL?

PEOPLE HAVE BEEN trying to control nature for thousands of years. Why, exactly, is our desire for dominion over the natural world so strong? No other species intentionally shapes its environment to fit its needs like we do. There are certainly animals whose actions constantly improve their habitat in their favor. Many large herbivores—elephants or deer, for example—require open plains with scattered stands of trees. Thick forests do not offer them sufficient quantities of the grass and the other green plants they like to eat. Through constant browsing, mostly by grazing on trees, they make sure dense forests don't take over. And in places where the tree canopy does manage to close, they damage trees so severely by stripping off bark that trees are constantly dying. This bark removal, however, happens not because the animals are intentionally creating new feeding grounds but because they are hungry. It is only incidentally that they are improving their chances of having more food to eat in the future.

We find something similar in our human past. Nomadic hunter-gatherers surely did not mean to transform large stretches of the landscape, but nevertheless, they left their mark on animal populations and chopped down trees here and there to make fires and build tools. Most of the forest, however, was left untouched—at least until the arrival of agriculture. Then, people started cutting down forests for livestock grazing and plowing fields to cultivate crops. The natural world around the small settlements of the Stone Age underwent massive changes. And yet even then the changes were kept within bounds and the land was still mostly covered with primeval forests.

THE DEVELOPMENT THAT changed everything was the establishment of nation-states. When you are able to subject large numbers of people to the same rules over a wide area, division of labor becomes possible on an unprecedented scale. This system eventually led to the cars and smartphones we have today, although most people are incapable of producing even one of the components that make up these staples of modern life.

The Ancient Egyptians founded the nation-state over five thousand years ago. One of the things it enabled the pharaohs to do was to build their immense pyramids. The largest of these, the Great Pyramid of Giza, which was built by the pharaoh Cheops, required 2.3 million stone blocks, each weighing more than one metric ton. To have the pyramid completed in his twenty-year reign required the chiseling, transportation, and placement of one block every two minutes.[80]

A nation-state can exploit nature much more efficiently than a small clan is able to do. The state can organize and redistribute people over great distances. And that happened again and again all over the world. Whether it was the Aztecs, the Chinese, or the Romans, they all spread over large areas and began to shape the landscape to their needs. The better they became at controlling nature, the more reliably they could plan and the more efficiently they could generate products.

WE CONTINUE PERFECTING ways to control nature to this day. We have done such a good job in the places where we have settled that nature there is almost unrecognizable. The founding of the nation-state was the starting pistol in a race to keep nature in check, a race that has hopefully run its course because of what we now know. This distancing from nature is most obvious in cities. No one in the city talks about forests, even though there are plenty of trees around. The bond that connects us to nature, however, has never been totally severed, so it is hardly surprising that the more we distance ourselves from nature, the more our love for it is awakened.

21

OUR LONGING FOR AN INTACT WORLD

IS THE TREND toward nature, and especially toward the forest, a movement we should embrace as unequivocally positive? Might it perhaps indicate an increasing flight from reality by a growing number of people who don't want to hear about politics or environmental destruction in their leisure time, spending it instead in search of an intact natural world that exists only in their imaginations?

I've often heard people level this criticism at my bestselling book about trees. They align it with works of fiction as though it were a detective novel. That in itself is not a negative judgment—who doesn't enjoy reading a good story about tracking down what was really going on?—but people who read the book this way miss out on the opportunity to engage seriously with one of the most numerous lifeforms that surrounds us. Instead, reading my book becomes another escape from today's world, which can also happen when you play video games, watch soap operas—or read other, similar nature books.

IT'S NATURAL TO want to slow down and relax after a stressful day. Whether we play a sport, eat delicious food, or read a good book, recovering from work by doing something different is completely normal and something people have been doing in various forms since we appeared on this planet. That's precisely why we created cave paintings and invented artifacts such as musical instruments.

In modern times, we are now often—or sometimes exclusively—surrounded by the trappings of civilization. Our homes, vehicles, thoroughfares, and workplaces—aren't they all artificial constructs made from unnatural materials giving off smells that have nothing to do with the earthy aromas of forest and field? Civilization is, after all, the opposite of nature and encompasses everything we have made. In the most general terms, it includes farmed fields and even forests of planted conifers. But if our daily lives are spent exclusively in places where nature isn't present, doesn't it become a necessity for us to seek out our natural environment every once in a while? Might we not have an instinctive longing for the surroundings our senses were made for? You could see the current back-to-nature movement as breaking out of a self-imposed prison cell rather than a fanciful, laughable rejection of reality.

IT IS IN exactly this negative sense that critics couple the term "escapism" with the term "esotericism" to describe our rekindled love of the woods as something that has no basis in reality—often encoded in snappy headlines such as "Is the Tree the Better Person?"

Under this headline in the Swiss newspaper *Tages-Anzeiger*, Martin Ebel wrote that the idea of the forest as the perfect example of nature and as the opposite of cities and industry is a German invention. The Romantics, he wrote, were the ones who elevated wild nature to the realm of the spiritual.[81] In the article, he ascribed to Westerners a particular perception of the forest, without, it must be said, disputing any of the scientific discoveries about the effect the forest has on our bodies and minds. I see Ebel as a typical defender of the old view of nature as a big machine, a soulless system, where environmentalists are now seeking salvation.

And he has a point, at least with respect to the changing views of the forest in Europe. It was the Romantics who gave the forest back its positive image. In Schönbuch, a large forested area in Baden-Württemberg, I got to visit a place called Olga's Grove. In the stifling heat of July, we were filming a segment for my television program and had been walking for hours through severely mistreated forests (or, I should say, plantations). Deciduous woodlands with little trees that must have been only a few decades old but had still been thinned were followed by stands of pines with large clearings full of fresh stumps. The logs piled along the roads reminded me of stranded whales.

After a steep climb, the scenery changed abruptly. We arrived in an old beech forest where the air was noticeably cooler. Narrow stepping stone paths wound their way through the trees. Here and there, benches invited you to pause a while, and small pools reflected the few rays of the sun that penetrated the crowns of the mighty trees. I imagined this forest must be natural, at least to some extent. But Henning, the producer, brought me back to Earth. There was nothing natural about it. On the contrary, the grove of beech trees was part of a park built for the enjoyment of a romantically inclined Russian grand duchess called Olga. Her husband, King Charles I of Württemberg, planted the grove on the hillside in 1871. Just a few decades later, the park was reclaimed by nature and forgotten. It wasn't until the 1970s that the local forest agency carefully restored the pleasure grounds, leaving the stand of trees basically untouched. That is why today it looks like an old forest with groomed trails running through it.

Back to the Romantics. Was that when we began to appreciate forests? Earlier, did we consider them to be dark places where terrible things happened, just as the Brothers Grimm always depicted them in their fairy tales? And, more importantly, it is only Westerners who yearn for the forest? Along with the terms "esotericism" and "escapism," the word "decadent" has wormed its way into my head. It is the third in the triad of terms used to discredit our new sensibility for nature.

IN A WORLD where almost all of our material needs are met, are we now really decadent, over-satiated, and simply seeking an outlet for our weariness with life? An oft-repeated argument that embraces this point of view and defends managed conifer plantations while rejecting forest preserves where trees are left to grow old goes like this: if we want to protect forests to keep our own environment intact and not allow any trees to be cut down, then our need for wood products must be satisfied by importing more wood. That, however, would mean more rapacious logging in rainforests. Therefore, it would be better to open up all our native forests to the timber industry.

I find the real decadence lies in arguments such as this. People who, bolstered by such arguments, want to plunder natural resources while accepting the decline of standards (such as the sustainability of ecosystems) and threaten to plunder other places if they do not get their way are only demonstrating their ignorance. Contrary to what they are arguing, protecting more forests than we have up until now is a question of both material and cultural survival. And what can protect forests more effectively than people having positive feelings about them? All the Romantics did was to recognize this once again, because trees were seen very differently before the darkness of the Middle Ages, as the ancient Celtic and Germanic tree religions demonstrate.

I see the growing attraction of nature, and of forests in particular, as at least a temporary retreat from our human-made urban world into the ecosystem for which we were created and on which we are, ultimately, still completely dependent. Cities, after all, are nothing more than a concentration of our products, products that we trade here but generally don't make here and that certainly don't grow in between the buildings. The artificial world of the city offers all kinds of stimuli for which we were not originally designed.

ONE OF THESE stimuli is noise. In 2016, the German Federal Environment Agency conducted a survey of a representative sample of people asking about the main sources of sound pollution in their lives. Street noises came in first. Number two, surprisingly enough, was noise from neighbors. Then came industry, airplanes, and trains. Often noises came from multiple sources at the same time.[82]

Noises cause cardiovascular diseases. The World Health Organization therefore recommends that long-term nighttime sound pollution not exceed a decibel reading of 40 dB(A).[83] That means that anything louder than a soft whisper can cause sleep disorders if it persists over time. Is spending the night outside a better option? Not necessarily—it isn't completely quiet in the forest, either. Trees rustle in the breeze, birds sing, and deer bellow. A light rain is as loud as 40dB(A). A storm is more than 80 dB(A), which is close to the decibel reading for a jackhammer. The difference is that unlike city sounds, sounds in the forest are not there all the time, especially not the loud ones.

If you're up for a small adventure in the forest at night, you can test this for yourself. All you have to do is take a walk after dark. It's perfectly legal, and you won't disturb the animals as long as you make some noise. If you don't make any noise, the animals will mistake you for a hunter and that will stress them out.

You won't hear any of the eerie noises people often talk about, because although the forest at night is not completely quiet, it's usually quite peaceful. The breeze drops as soon as the sun disappears below the horizon. Most of the animals fall silent. The most you'll hear is a trembling *hoo-hoo-hoo* from a lonely owl. In the forest lodge where Miriam and I live, it is so quiet that some guests don't sleep well. They miss the sound of the streetcar or tires rolling over asphalt under their bedroom window. When it's raining, however, sleeping under trees is pure relaxation for both your ears and your circulation.

APART FROM SEEKING respite from noise, many people go into the forest for the air, which is said to contain particularly high levels of oxygen. Doesn't a walk under the trees, in comparison, say, with a walk in the city, provide the purest shower of oxygen for our lungs? That is not always the case. To explain why this is so, let's take a look at winter. At that time of year, the trees have lost all their leaves and even the conifers are hibernating. That means that in this season beeches and other species are living off the reserves of sugar they built up and stored over the summer. When this sugar was being created, the net effect was that the trees released oxygen, but in winter the process is reversed. The trees burn sugar in their cells and, just like we do, they breathe out carbon dioxide. Don't worry, though, it would take you a long time to suffocate, because a large part of the oxygen in the air comes from the ocean and the supply is constantly topped up by the winds that blow in our direction.

One special form of oxygen is ozone. (It is made up of three molecules of oxygen.) Paradoxically, when it comes to ozone, you're better off in the city than in the country. Ozone is an aggressive, toxic gas that damages the lungs. This is why the Federal Environment Agency in Germany issues special warnings when levels rise steeply in the summer. One way ozone is formed is when carbon monoxide is released in vehicle exhaust. When the sun shines brightly, as it does on a hot summer's day, the exhaust reacts with oxygen in the air. Ozone produced in the city stays in the city and immediately binds with new exhaust emissions, which means it is broken down. If the wind blows this gassy cocktail out into the countryside, however, it is set free and accumulates in the air. On hot summer days, therefore, the amount you breathe in becomes problematic mostly in rural areas. And so, even if our instincts and desires lead us back into nature, we've manipulated our surroundings so much that things are not always in balance out there. With our next topic—dust—things are, however, much better in the forest.

SINCE THE SCANDAL about diesel fuel in the cars and trucks we drive, controlling air pollution is once again at the top of the German government's agenda. According to a report released by the European Environment Agency at the end of 2018, air pollution was responsible for about 442,000 premature deaths in Europe.[84] The reason is not only particulate matter in the air, but also nitrous gases. These pollutants come from exhaust pipes, which have been justifiably singled out, as well as all the chimneys out there. These chimneys are evidence of the more than 12 million woodstoves the Federal Environment Agency says are smoking away throughout Germany. In the United States, as I mentioned earlier, 30 million people live in homes where wood is burned to provide heat, so there are many chimneys adding smoke to the air there, as well.

I'm intentionally being a bit negative here, because a lot of people don't know that heating correctly with wood is an art. It all starts when you light the fire. Contrary to popular opinion, the starter used to ignite the kindling should not be put under the pile of wood in the combustion chamber but on top. Only then do you get a clean burn that doesn't smoke out the entire neighborhood. Also, the wood must be really dry if it is to burn fairly cleanly. Fairly cleanly—that's a relative term. While the whole world is focused on diesel emissions, wood burning puts more particulate matter into the air than all the cars and trucks combined.[85] To be clear here: a wood-burning fire is a fine thing, and I enjoy heating with our tiled stove. In Germany, the question is whether we want the federal government to subsidize this method of producing energy as it is still doing with pellet stoves.

In the United States, there are laws and standards that apply here. In some parts of California, for example, wood-burning fireplaces are not allowed in new constructions, and many states have emissions standards for appliances that burn solid fuels such as wood pellets or coal. Another factor to consider is agriculture, where emissions from liquid manure spread on fields contribute hugely to the problem of particulates in the air.[86]

We are lucky that we have the forest. If absolutely necessary, it can filter up to 20,300 tons of dust from the air per square mile every year (7,000 metric tons per square kilometer per year).[87] In this respect, forest air really is clean.

IF YOU ADD the effects of the forest and nature I've described here to what I've explained in previous chapters, is it any wonder that nature constantly calls us out of the cities and back to our roots? Isn't this a healthy instinct, which shows that we are still completely in touch with our senses? The people who call this escapism are the ones who are out of touch.

22

LEARNING FROM CHILDREN

GONE ARE THE days when you could just get up and go out into the forest. These outings now take meticulous planning. And I don't mean just setting a date, but also planning exactly how the day will unfold. Let's take a typical example. You leave at 9:00 in the morning to get to the parking lot by about 9:30. Because you want to arrive at the restaurant around noon (a table has been reserved) and you've been late leaving the parking lot, you have to make good time on the trail—heaven forbid your reserved table gets given away to someone else! Everything turns out well, thank goodness, and after a slap-up meal, you return to your car at a more leisurely pace. The only thing that was annoying was the children. They kept dawdling, they complained about having to hurry up, and sometimes they simply refused to move. Every branch was interesting. Every mossy stump was worth exploring.

We should follow the children's example. After all, the reason we're out in the forest in the first place is to enjoy it. And yet, because of our hectic daily schedules, which we can now plan to the minute on our cell phone calendars, we carry our busyness over to our leisure time. And to our children.

NOT LONG AGO, I was in the forest with a journalist who wanted to write a piece on the tours I lead for children. She brought along a friend with two young ones so she could observe how I interacted with groups. She was not the only one to learn things that day. I, too, learned a lot from the encounter.

The children were all fired up at first. They wanted to uncover all the secrets of the forest—right now! No sooner had I shown them something than they asked, "What's next?" I found myself thinking about the way we consume entertainment in the internet age. We are constantly flipping from one device to another, from television to tablet, for example, and watching both at the same time. I imagine that such behavior rubs off even on small children.

These two little ones began to behave differently after about half an hour. They started to walk more slowly and then stopped completely at a stump. Here I showed them all the things they could find under the bark. There were sow bugs that quickly hid in the nearest crack as soon as they were exposed to bright sunlight, centipedes that snaked their way over the rotten wood as they hunted down their insect prey, and ants that had chewed themselves a nest in the stump. After twenty minutes, the adults were ready to move on, but I waved them away. The children were still so caught up in the excitement of discovery that I didn't want to interrupt them. I realized that this was exactly what an ideal walk in the woods with children should look like—with them dictating the pace.

If children have to keep up with the adults all the time, if they have to keep quiet so as not to disturb the animals, and if at every discovery they are told to please get a move on, it won't be long before a walk in the woods will bore them to death. Even I found these walks boring when I was young. A weekend excursion always had a destination that had to be reached by the shortest route with as few interruptions as possible. My parents, no doubt, felt very differently about these outings, as they chatted away while walking along. For me and my siblings, however, the excursions were much less interesting than when we were out and about with our friends, building forts, playing cops and robbers, and, most importantly, making as much noise as we wanted.

ON FAMILY OUTINGS, parents often caution their children to be quiet so as not to disturb the animals. That is, however, completely unnecessary. When animals hear people making noise, instead of getting stressed, they relax, because they realize right away that noisy people have no intention of hunting them. Hunters, after all, are silent when stalking their prey. Something happened recently when my colleague Josef Eichler was leading a forest tour: the group listening intently to what he had to say was, it turned out, not his only audience. A deer was calmly standing right behind them, listening equally as intently. It had its eye on the potentially dangerous people, but clearly recognized that they were harmless.

When I give tours for children, the first thing I do is to tell them to yell as loudly as they can. That helps them overcome any initial shyness (after all, they don't know me) and calms them down. It also calms the animals down, because they realize these little forest visitors aren't a threat.

Then we usually get to the next main issue: dirt. The rule on my tours is that the children can get as dirty as they like. This rule actually applies only to their clothes. Hands are a different matter. It's too much to ask most parents to silently stand by and watch their child unpack a lunchtime sandwich with dirty fingers. But is what we find on the forest floor really dirt? Of course not. It's mineral-based soil combined with bits of humus—neither of which is poisonous or unhygienic. So, don't worry. Let your children dig around with their hands. Better yet, join them. Children are wonderful teachers because they have no inhibitions about getting close to nature. If we manage to enjoy the forest as much as they do, perhaps that will help reinforce our tie to nature, which has become slightly worn of late.

23

THE PARADOX OF CITY AND COUNTRY LIVING

I REMEMBER MY FIRST year in the forest I manage very clearly. Rain or shine, I loved being out in nature and every encounter with a wild animal was thrilling. But every once in a while, I was also horrified. Whenever I crossed a creek running through a steep valley near one of the small villages, I found garbage. Old glass bottles, car batteries, or even parts of complete cars peeking up from the forest floor. What was going on? There were also mounds of plastic detergent bottles and pesticide containers. What kind of environmental miscreant had thrown all this away here? After a while, I realized that these were old garbage dumps. I had no idea that garbage pickup was first introduced in rural Germany in the 1970s.

People don't want broken glass or jagged metal in their gardens or fields. That would be a hazard when working the land. It would be dangerous, and the glass and metal might even end up in the hay fed to livestock. The solution was simple. Get rid of it in the forest. People didn't like the forests much anyway, because they had been planted in the nineteenth and twentieth centuries against the wishes of the local residents by authorities pushing economic development.

SINCE THEN, HOWEVER, the forests have brought employment and a certain level of prosperity to the surrounding villages. Even back in the 1950s, men traditionally worked their small farms in the summer and then, in winter, when there wasn't much to do on the farms, they hired themselves out as laborers in the forest. Unregulated garbage dumps would have been a nuisance in their new places of work, as well. But what about the steep valleys where the creeks flowed? These gullies were not places where you could grow trees commercially, and they were easy places to get rid of your garbage. All you had to do was drive to the slope and tip the whole load over the edge—everything was now conveniently out of sight. In those days you didn't worry much about the environmental consequences of your actions.

Years ago, garbage consisted mostly of biodegradable materials: leather, wood, woven willow fence panels, cotton or wool fabric—and none of that presented a big problem. Glass bottles could usually be returned for their deposit and so in general the only non-biodegradable items that were thrown away were old plates and stoneware storage jars. Things changed after the Second World War. Containers were increasingly manufactured out of plastic and a flood of disposable packaging found its way into homes—and from there into the forest.

This form of disposal basically stopped in the 1970s, but no one wanted to haul away the garbage already piled up in the forests behind the villages. The solution was simple: dump soil over everything. And this was exactly what happened. Unfortunately, it was done in such a slipshod fashion that after just a few years, when the fill had settled, bits and pieces began to poke above the surface. To this day, unknown numbers of old garbage piles lie dormant near every village. And to ensure the mayors can sleep better at night, the municipalities have negotiated insurance policies that will take effect if this legacy from earlier times ever causes significant environmental damage.

I'M MENTIONING THIS environmental disgrace because the tradition of tipping garbage down slopes continues to this day. Everything the local population deems harmless still gets thrown into the nearest creek valley. Usually the material that lands on top of these sensitive wetland areas is compostable green waste with a few non-compostable items like plastic plant pots or rubble from construction sites mixed in. This despite the fact that yard waste containers to collect lawn and hedge trimmings have been available everywhere for some time now. The waste is then taken to a central location in the district to be processed into compost. Plumes of black smoke are also a regular sight near homes in isolated areas outside the villages—a sign that people here still burn their garbage. Although most village residents find this inappropriate, no one intervenes.

Even in what to us was faraway Sweden, my family and I encountered similar practices at the start of one of our wilderness vacations. We had rented a horse and covered wagon for a week. We wanted a relaxing drive through the Swedish forest at a leisurely pace: sleeping under the stars, eating, and simply enjoying nature. We left Germany and when we arrived at our destination, we turned up the driveway that led to the farm yard. A red painted house awaited us, surrounded by weathered barns. The owner-operator was in the process of constructing a new barbecue area, and he was smoothing out the soil fill with a rake. Unfortunately, he was not quite ready for our arrival, and so we noticed that he was hiding garbage under the fill—quite a lot of garbage. We don't have to look as far the Third World to find fault with the rough-and-ready way people deal with garbage and nature, even if the amount of pollution is far greater there.

DESPITE THESE PRACTICES, country people dearly love the places where they live. It's just that this love is a bit rough around the edges. The garbage example shows in stark practical terms that thinly populated areas have developed a different way of dealing with nature, a way that certainly would not have been possible in the city.

This impacts trees, as well. If a tree is in the way in the country, down it comes. Permit? Who needs one? It's a completely different situation in the city. In Germany, if you cut down a tree without a permit, you can be slapped with a fine of up to €50,000 (which is about US$55,000 at today's exchange rate),[88] and many cities in North America have tree ordinances. You might argue that cities have tree protection ordinances and rural areas don't. But that's the point. Cities and the councils that represent them believe trees are so important that they fight for each and every one of them. Even if in practice it's often the municipal authorities themselves that hack away at trees and bring up safety concerns—when it comes down to it, cities pay more attention.

That's usually not the case in the countryside, and forestry is a good example. A single tree doesn't count for much when, in the average forest parcel, you have to cut down ten thousand to twenty thousand trees each year. But that's not the worst thing. In the first part of summer, there are countless birds' nests up in the branches. Yet, even in bird sanctuaries trees can be cut down when birds are sitting on their nests, causing the deaths of hundreds of thousands of baby birds.

While foresters shrug their shoulders and point to timber contracts that must be fulfilled quickly, homeowners are fined even if their offenses cause little harm. In Germany, for instance, federal environmental law even forbids them from trimming their hedges between March 1 and September 30. Forestry, with its huge nest-destroying machinery, is given a pass. Commercial forestry is allowed to be a little rougher, an attitude that is no longer appropriate for the times in which we live.

APPROPRIATE FOR THE times. Those are key words for the forest, which really should be thought of as timeless, because an untouched forest changes slowly over the course of thousands of years. People bound up in commercial systems, however, live shorter lives and impose the preferences of the day on ecosystems. When it comes to trees, with their long lifespans, these preferences can still be seen decades later. Forests act as a mirror to our cultural past.

24

TREES, TOO, ARE FOLLOWERS OF FASHION

WAIT A MOMENT. The chapter heading is somewhat deceptive, because trees themselves know nothing of fashion. It's people who love change, but only when everyone else is doing it. As soon as a trend gets going, most of us get on the bandwagon (me included).

Plants don't escape these trends. You might have noticed this with fruits and vegetables. A new superfood is always popping up, a plant said to have particularly health-enhancing characteristics. Take the goji berry. The name it is sold under sounds far grander than its common name: wolfberry. The bush probably originated in China. In Europe, it is a recently introduced, non-native species, and it's spreading. It's becoming increasingly popular in North America, as well. The small, orangey-red fruit—about 0.75 inch (2 centimeters) long—are apparently stuffed full of things you need:

antioxidants, essential fatty acids, iron, various vitamins. It sounds too good to be true. In addition, its pleasant taste is touted as a welcome addition to your morning bowl of cereal. At least that's the theory. Because Miriam and I like to experiment, we ordered two plants. We waited with bated breath for the flowers and then the fruit. After two years it was finally time to take a little taste. The bland sweetness of our first bite quickly turned unpleasantly bitter. No matter how healthy goji berries are supposed to be, we'd rather leave ours for the birds.

THERE ARE ALSO trends for houseplants. From cacti to yucca and peach sage, a variety of popular plants can be grown inside and carried out into the yard, as well. Why should things be any different in the forest? Or I should say in commercial forests, because most forests, in Germany at least, are a collection of trees planted by people. Everywhere foresters plant trees, current fashions play an important role. Forest agencies, of course, characterize what they are doing slightly differently: they are committed to following the latest science.

But in truth the custodians in green succumb to the same temptations as normal shoppers. What draws them in are exotic trees. If an exotic tree is going to grow in their forest parcel, it must come from a similar climate zone and, most importantly, tolerate relatively cold winters. For German foresters, that describes many forests in northern latitudes—North America, Europe, and Asia. That's why you find so many giant sequoias here. I saw one when I was visiting Olga's Grove in Schönbuch, that large wooded area in Baden-Württemberg I mentioned earlier. The sequoia is almost 165 feet (50 meters) tall and has a diameter of 6 feet (1.8 meters) at chest height. That's impressive because most of our native trees in Germany—the beech, for instance—run out of steam after growing to 130 feet (40 meters).

But the sequoia stands out in the forest like an elephant among deer. Perhaps a comparison with a zoo works even better. These individual plant trophies among the oaks and beeches grow alone, without the companionship of their fellows, deprived, even, of their own ecosystems. A complete, well-functioning forest includes thousands of species networked together in a delicate balance. A lone tree, even if it is the largest tree around, doesn't give you any sense of what it's like to be in a North American forest.

People have always been fascinated by fast growth. In the 1960s, foresters were mad about poplars. Balsam poplars were bred and cross-bred until the trees shot up to a height of 100 feet (30 meters) in just twenty years. Compare this with spruce, which are known for their fast growth but grow little taller than 30 feet (10 meters) in the same amount of time. What people completely forgot to consider was what would happen to the trees once they were grown.

A major customer, a maker of matches, dropped out of the market when disposable lighters came on the scene. A similar thing happened to companies that made wooden fruit and vegetable crates from chipboard. The now thick and sturdy trunks of the mature poplars became a liability. No one wanted their wood, and the young giants turned out to be hazardous along roadsides and trails. Their upper branches are as brittle as glass and break with just a little wind or snow. The upshot was that poplars were taken out everywhere, and their wood was sold at a loss. In both Germany and the United States, poplars are currently experiencing a renaissance in short-rotation plantations, where, after a few years, the spindly trees are harvested, shredded, and burned in biomass power plants. But because this is tree farming and has nothing to do with forests, I won't delve into the subject any further.

ANOTHER EXAMPLE OF a fashionable tree is the grand fir or *Abies grandis*. Its home is a relatively small area along the northwest coast of North America, where grand firs grow in mixed coniferous forests. Grand firs meet many of the criteria that attract foresters. They grow quickly, much more quickly than the spruce that have dominated German forests thus far, adding up to 3 feet (1 meter) to their trunks every year. Unlike Douglas-fir, they tolerate periods of drought well, which is important because water shortages will become more common as a result of climate change. These trees even seem to have little difficulty weathering storms. They grow their roots very deeply, so they are anchored more firmly in the ground than spruce, pines, or Douglas-firs. They have been grown in Europe since the nineteenth century, but their breakthrough moment came after the arrival of Kyrill, a cyclone that hit Europe in 2007. Numerous plantings are now awaiting their eventual fate. But what will that be?

I had a few grand firs in the forest I manage. A forester keen on experimenting had mixed them in with spruce decades ago. Since then, they had grown into stately trees, and I had them cut down as part of our regular harvest. The saw mill owner, who was buying them to make construction lumber, did not think much of them, and neither did most of his colleagues. The soft wood is not as high-quality as spruce and so is correspondingly cheaper. That was why, up until Kyrill, no one set great store by this tree. The grand fir was dusted off as people discussed what action to take in the face of climate change. No one wanted to do without conifers, and because the future of spruce, pines, and Douglas-firs did not look good, it seemed then, and still seems today, that grand firs will be the saving of coniferous forests in Germany.

WHY DO FORESTERS let themselves be beguiled by trends in trees? The answer is simple. Foresters are human. They love big trees, they love trees that are different, and they have fun with new developments. Above all, they like creating things. That culminates in a phrase I've heard from many of my colleagues: foresters "make forests." Without foresters, forests would be like sick patients with no chance of survival.

All of these attempts to make forests into what we want them to be reminds me of rearranging living-room furniture. Trees as decorative sofas and chairs—you can't get any further from nature than that. The workers in green, my profession, have lost contact with true nature even more completely than people who live in the city.

25

THE LONG, HARD ROAD BACK

EVEN WHEN I was a young boy, I wanted to protect nature. After I left school, the profession of forester—who I thought of as being a kind of guardian of trees—seemed exactly the right fit for me.

This career still has a dash of romanticism about it. Stories and fairy tales have contributed to this aura, including the 1954 German film *Der Förster vom Silberwald* ("The Forester of the Silver Wood"), in which Rudolf Lenz, the newly arrived community forester, prevents the clear-cutting of a mountain slope. As I've mentioned before, most foresters—influenced by the way their profession is viewed—are convinced their actions benefit the forest. If we are to keep forests, we must care for them or everything will fall apart. I was indoctrinated with this point of view during my studies. I don't think there was any ill intent. Quite the opposite, in fact. Foresters love forests and want to preserve them.

In these times of climate change, trees are too slow to react as their environments are altered. For instance, trees in warmer climes cannot migrate north quickly enough, and native species can't escape even farther north. Trees need hundreds if not thousands of years to migrate—time that neither they, nor we, have. Therefore, jumping in to support the trees and helping a bit by mailing seeds and raising trees in nurseries seems like a really good idea. Besides, we still need to produce wood and, if possible, in a way that supports the local timber industry. It's not easy to master this balancing act—especially when state forest agencies are supposed to be managed so they turn a profit. This tricky combination achieves exactly one thing: frustration. If you try to calculate the benefits, you find the forest is not nearly well-enough researched for us to be able to assess how well our manipulations of any species work.

Here's an example from where I live. Environmentalists are monitoring the decline of species, and some of these, such as grasshoppers, red wood ants, and wild bees, thrive where it's sunny and warm. You can help these sun lovers by opening up clearings in the forest. When more sun hits the ground, more grasses and herbaceous plants grow, so the species I've just mentioned can find more food. Furthermore, these insects need direct sunlight to warm them up so they can function. Consequently, foresters clear more trees for conservation of these species than is healthy for the forest. Many sun-loving species did not live in the forest originally; for them it is a substitute habitat that replaces the open pastures they have lost. Conventional farmers spray to kill everything that doesn't serve their production goals. They are the ones who should be adjusting their farming practices.

Instead, the forest agencies give in to pressure and offer the displaced pasture residents a home.

BUT THEIR NEW home was already occupied, as the records for beech woods in Germany clearly demonstrate: about ten thousand species of animals have been identified as depending in a multitude of ways on the cool, dark, and damp conditions under the giant trees. Many of these species are at a disadvantage because they are very small and some of them are also ugly. Mites, for example, definitely have fewer defenders than hares, former plains dwellers that now feel quite at home in many forests.

This is the situation in one of the nature preserves in the Hunsrück, a low range of mountains just to the south of where I live. I could not help but admire an old pastoral forest: a grassy landscape dotted with ancient oaks. In earlier centuries, people grazed their cattle in the pasture and drove their pigs out to forage for acorns under the trees in fall so they could lay on a thick layer of fat before they were slaughtered.

Forest animals don't feel comfortable here, but butterflies and deer like it very much. And so, every year, a lot of money and effort goes into mowing with big machines to cut off the tree seedlings that would help the forest get established once again if they were left to grow. Deer, for example, would certainly prefer to eat grass in the meadows along rivers. In their natural state, these meadows contain a mix of trees and grass. The reason for the fairly open landscape used to be ice floating down the rivers. Frozen rivers broke up in spring and, as the snow melted, the swollen rivers carried thick, heavy ice floes through the forested river meadows, shaving away young trees and severely damaging old ones. In the ensuing gaps, grasses and low-growing leafy plants grew, subsistence foods for large herbivores. Aurochs, moose, and wild horses are long gone from Europe, but up until recently deer have been able to find sanctuary in these areas. It would be lovely, however, if we could return to them at least a small part of their original territory.

But river valleys are where we like to live, and we take up a lot of room. Every day, asphalt and concrete make further inroads into the river flats. In Germany, we pave over an additional 38 square miles (100 square kilometers) or so a year—that's the size of an entire national park in the Black Forest, the Eifel, or the Hunsrück. But there's no room for national parks in these concrete jungles, and people don't want them there.

Close by, down in the valley, are the most productive soils. They lie in areas that used to be subject to frequent flooding and so are particularly rich in nutrients. They would be especially well suited for food production for both people and animals, and yet they are now buried under roads and houses for the foreseeable future. That means that the animals of these grassy landscapes have been pushed back into the forest, where they have difficulty finding the kind of food they need. Foresters can help these animals by creating clearings in the forest, where all kinds of grasses and low-growing leafy plants can move in among the trees. At the same time, acres of land are being completely cleared, then leveled and seeded, to create so-called browsing (which means feeding) areas. Deer are not the only ones drawn to these areas; hares and butterflies like them, too. In today's intensive commercial forestry, many German forests are looking more like open grassy plains than deep woods.

In the United States, battles are constantly being fought to protect old-growth forests squeezed by urban growth. In the Oakland Forest near Providence, Rhode Island, a small stand of centuries-old maple, beech, and tulip poplar, once part of Cornelius Vanderbilt's country estate, was saved from a developer at the last moment by a local group. Citizens in Memphis, Tennessee, stopped a highway from cutting through Overton Park's ancient trees, part of an ecosystem that dates back at least ten thousand years. The city of San Francisco has within its limits the old Oak Woodlands in Golden Gate Park, an area that has escaped the ornamental landscaping installed elsewhere in the park. These small fragments are a vivid reminder of how dramatically land use around the world is changing and how much pressure is being put on forests.

NOT EVERYONE WANTS to be a woodland revolutionary, but it is actually quite simple to do something for the trees: just use less wood. It might sound odd for me to write that, because as you are reading this sentence, you are, after all, holding a book in your hands—made of paper. And paper is made of wood. As is furniture, roofing timbers, or backyard fences. What they all have in common, however, is that they all last for a long time. The better the quality, the longer they last and, therefore, the less often they need to be replaced—and that saves wood.

The same cannot be said, however, for packing materials. As part of the worldwide effort to keep plastic out of the environment, more and more paper and cardboard is being used instead. That spares the oceans, where a huge amount of plastic ends up, but it is hard on forests. Even today forests cannot keep up with our immense demand for wood products, a demand that is steadily rising.

The main cause is that wood is considered an environmentally friendly raw material for two obvious reasons. First, it is renewable. When a tree is cut down, another can always grow, provided that the land has not been commandeered for agriculture or urban growth. Second, wood is thought of as carbon-neutral when it is put to use. After all, when a tree is burned in a fireplace or stove (and all wood ends up either there or in a biomass power plant after its useful life is over), it cannot contribute more greenhouse gas to the atmosphere than it captured and stored when it was growing.

Let's stick with this idea for a while. Using wood is not carbon-neutral. The calculation with the tree itself is correct. When it is used and burned, it is impossible for it to release more carbon dioxide than it stored in its wood as a byproduct of photosynthesis when it was growing and alive. But this is only one part of the process. Leaves, branches, bark, fruit, dead trees—all accumulate in the soil in the form of humus. Moreover, untouched forests contain at least twice the amount of living biomass as managed forests.

When trees are cut down, this reservoir is emptied not once but twice. First, the amount of living biomass is reduced, and second, the humus in the forest floor is depleted. This happens because sunlight penetrates and warms the ground. Fungi and bacteria shift into high gear and begin to consume the organic components of the soil until they are almost all gone. As part of the decomposition process, carbon dioxide is released into the air, just as it is when we digest food. All these processes are at work in managed forests, making burning wood as detrimental to the environment as burning oil or coal. Wood, therefore, is not the untarnished, environmentally friendly raw material it is made out to be.

Paper has the environmental advantage that it can be recycled without leaving residues, as long as it is print-free. But how many bags out there are free of business logos or synthetic materials? To say nothing of the fact that such carrier bags are seldom used more than once because they are not particularly durable and tend to rip easily.

And what about that first argument about renewable resources? Things look somewhat grim here, also because the demand for wood products worldwide has grown so much that it cannot be satisfied through sustainable forestry practices. One ancient forest after another is being cut down and replaced with barren monocultures of eucalyptus or pine. Our laudable intent to save the environment from drowning under a flood of plastic by using paper instead is unfortunately fueling destruction of a different kind.

SO, WHAT IS the alternative? It has to be this: less packaging. We had a similar movement in Germany back in the 1970s and '80s when the slogan was "burlap instead of plastic." Until the mid-1990s, multiuse containers were widely available. You brought them with you and filled them with milk, processed meats, and cheese. As I mentioned in an earlier chapter, I still remember campaigns at school where we spent years collecting the foil lids of yogurt containers. After all, they were made from valuable aluminum that was energy intensive to mine and process. To make this clear to us, my chemistry class was once taken on a field trip to an aluminum plant. We were deeply impressed when the high-voltage electric currents made the coins in our wallets stick together.

After that we had no doubt that a lot of energy was being used in the production process, and we redoubled our efforts to collect every scrap of aluminum foil we could. Given the massive aluminum rims sported, it seems, by every second car these days or the number of aluminum wheels on bicycles, such efforts from the past seem quaint and not particularly effective. And they didn't last very long.

With the fall of the Iron Curtain, international policies of détente, and economic growth not only in Europe and North America but also in emerging and developing countries, environmental protection wasn't completely forgotten, but it was somehow not as sexy as it once had been. I noticed this back then when I took young people out on tours of my forest.

In the early years, my young audiences were still well informed and engaged, but by the early 2000s, their knowledge and enthusiasm had definitely waned. It is only recently that the idea of environmental protection has once again become meaningful, and this time it seems to me it has much more staying power. One theme stands out above all the others: climate change. In this, trees are our natural allies, but only if we stop treating them as nothing more than biomass.

26

CONFRONTING CLIMATE CHANGE

WE'RE ALL FIXATED on carbon dioxide, transfixed like deer staring into headlights, unable to see anything else. And this means that other huge issues—water, for instance—are being overlooked. We all know evaporation cools things down. We benefit from this when we sweat in summer to ensure our body temperature doesn't rise too high. And forests do the same thing. Forests transpire enormous amounts of water—up to 130 gallons (500 liters) for a single beech on a hot summer's day. Transpiration reduces the temperature in the forest by many degrees, and you can feel this. It's not just the shade that causes the distinct difference when you walk from an open area into a forest; it's something the trees are actively doing.

Most species of trees don't like it hot; they prefer moist, cool conditions. At least, that's the case for trees in the temperate regions of northern latitudes, which is where the largest forests on Earth grow. When forests are cool and moist, less water evaporates from the ground and, in these conditions, trees photosynthesize quickly and efficiently. The well-known German meteorologist and television weather forecaster Sven Plöger once told me how aware he was of this effect. Temperatures would climb steeply in April as the sun rose higher in the sky, only to cool down again in May. Temperatures dropped as the trees unfurled their leaves and the new growth immediately began transpiring. It's amazing that the effect is so noticeable given that here in Germany only 12 percent of the original forest cover remains. And the remaining trees are not ancient trees, which are the ones that cool things down most effectively, but overwhelmingly extremely young commercial forests that don't manage water well. The residual cooling effect is nonetheless impressive and surely only a glimmer of what it once was and could be in the future.

CAN YOU IMAGINE the effect huge intact forests that cover large parts of continents could have on the Earth's climate? If you can, you're a huge step ahead of most politicians these days, because they want to fight climate change by cutting down forests. Burning wood is climate-neutral, we're told, because for every tree felled, a new one can be planted, and so there's an endless cycle of regeneration. Moreover, it doesn't make any difference whether at the end of its life a tree is left on the ground to rot with the help of bacteria and fungi or whether it's hauled off and burned. In both cases, it is turned into carbon dioxide that ends up back in the atmosphere. But dead rotting trees don't end up back in the atmosphere in gaseous form. Most of them end up in the ground as humus. There they continue to sequester the greenhouse gas as carbon for many thousands of years. Moreover, trees in ancient forests live much longer than trees in plantations, which means large amounts of carbon dioxide are sequestered in their living biomass.

There are completely different reasons, however, for why spruce and pine planted in orderly rows are not as good at sequestering carbon as ancient, intact forests. Even if they don't end up being clear-cut, nature often ensures their demise when the next hurricane blows them all over. Deciduous trees drop their leaves in winter, but conifers keep a thick coat of green needles on their branches, which reduces their ability to withstand high winds. When a tree is over 80 feet (25 meters) tall, the pressure on its roots when a storm tosses its crown back and forth can be so intense that the tree falls over. Once a tree falls to the ground, it is either left to rot or its wood is used (and later burned as wood waste). In both cases, all the carbon it contains is released.

All the carbon? Didn't I just explain how most of the dead wood left in the forest remains permanently in the ground in the form of humus? That is indeed the case, but it holds true only for ancient forests where an old tree dies every once in a while. The death of one tree doesn't change the microclimate, and the forest remains shady and cool.

Things are very different when it comes to clear-cuts or clearings created by storms. In these areas, the sun beats down mercilessly, which kicks fungi and bacteria into high gear. They break down all organic substances completely and make sure every last fragment of wood ends up back in the atmosphere as carbon dioxide. All the carbon dioxide in processed wood ends up back in the atmosphere, as well. An official in a large environmental association once told me that the average lifespan of long-lived wood products is just twelve years. After that, books, furniture, or construction lumber are waste products that are burned in incinerators, where every last bit of carbon dioxide they sequestered is once again set free. Together with the vast amounts of humus on the exposed forest floor that is broken down by fungi and bacteria, up to 260,000 tons of greenhouse gases are released into the atmosphere per square mile (100,000 tons per square kilometer). But that is not all.

The cooling effect I have just mentioned could be considerably more important than we think. If temperate zones in northerly latitudes, such as Central Europe, had been covered with native beech forests, then the extreme summers of the past few years certainly would not have been so intense, and it's unlikely that we would have watched thermometers top 86°F (30°C)..

Of course, this mental exercise is somewhat contradictory because in that scenario, there would be no modern industrial society fueling climate change. But even leaving industry out of the picture, we'd still be left with the impact of having nothing but remnant or replanted forests. Replanted forests, for starters, are not particularly cooling, which is partly due to the kinds of trees in them. The needles on spruce, for example, which are still the most commonly planted trees in Germany, are darker than the leaves on native beeches and oaks. And that fact alone means that spruce heat up more quickly.

In an international study, a team led by Dr. Kim Naudts from the Max Planck Institute for Meteorology discovered that the changes brought about by forestry practices in Europe over the last three hundred years have contributed to a rise in summer temperatures of more than 0.22°F (0.12°C), despite the industry's massive tree-planting programs.[89]

That might not sound like very much to you, but think about the vigorous debates going on today about whether the warming of the climate should be restricted to 2.7° or 3.6°F (1.5° or 2°C), even though the difference seems tiny. The number after the decimal point is extremely important, as the following example shows. If we assume an average rise in global temperatures of 2.7°F (1.5°C)—a level we will surely reach soon—then 0.22°F (0.12°C) is 8 percent of that total rise. That is larger than the contribution to total global greenhouse gas emissions made by a country such as India.[90]

A word of caution here: this example only works on a regional level. Whereas European forestry practices contribute to the rise in temperature only in Europe, thanks to global air currents, industrial greenhouse gases affect temperatures around the world. And yet the example, inadequate as it is, still points to an underlying truth. For residents, it matters how much the local climate heats up (and how much local initiatives can reduce the effects). How much local temperature changes can differ from global averages can be seen especially clearly in the Far North. There, in the most forested region of the world, climate change is hitting particularly hard. According to Professor Markus Rex of the Alfred Wegener Institute, temperatures in the Arctic are rising twice as fast as the global average.[91] That is leading to bizarre weather events, such as a reading of 43°F (6°C) in Greenland in February 2018, in the middle of the long, polar night—a reading 45°F (25°C) above normal for that time of year.[92]

FORESTS IN SIBERIA, Scandinavia, and North America, especially those that are near the transition zone to the tundra and therefore more adapted to extreme cold, are facing new challenges. Originally, summers here were short and mostly cool, and there was plenty of water to go around—trees don't transpire appreciable quantities of water when summers are short. The last snow melted in June and the first snow fell in September. In between, there was time for a little photosynthesis. Just enough to fuel very slow growth in the trees. Farther north there were no trees at all, or at least no trees of any significant size. When you have a growing season that lasts only a few weeks, all that can survive are grasses and a few small bushes. Botanically, even the trees here are classified as bushes because they rarely grow more than 10 inches (25 centimeters) tall. This is a landscape where dwarf birches and willows blend in seamlessly with lingonberries and lichens, and all of them spend most of the year tucked under a deep layer of snow.

Things are now changing. Climate change is causing the white cover to melt earlier and earlier, and it returns later and later after summer is over. The boundaries of the vegetation zones are shifting and the amount of precipitation is changing significantly depending on the region. Years ago, a reindeer herder told me that the amount of snowfall in her area was far greater than it used to be. That can mean starvation for reindeer, because they scratch the ground to dislodge the lichens they eat. The thicker the snow, the harder this is to do and the more energy it takes.

A SURPRISING EXAMPLE from Alaska shows how fundamentally the ecosystem is changing. There, beavers are expanding their range north. The reason: higher temperatures allow bushes and small trees to get established and grow taller. Beavers love this, because they use the branches and trunks of saplings as food and building materials. They pile up dams to make ponds where there was once dry land.[93] Activities that would make conservationists in Germany dance for joy, worry researchers in the Far North. The permafrost in the ground under the new bodies of water thaws out much more quickly. As this happens, the organic materials sequestered there begin to decay and release greenhouse gases into the atmosphere.

The only reason the beaver is making this northward migration is that the woody growth that provides its food and construction materials is making this migration, too. You could say that trees are, once again, conquering the Far North. From the perspective of climate, this has positive consequences. Trees capture carbon dioxide and store it as wood. It would, however, be better if we were to allow more true forests to thrive in lower latitudes. Then carbon dioxide would be sequestered where it belongs and the Arctic tundra could remain unchanged.

I'VE ALREADY MENTIONED how much people, foresters above all, enjoy shaping forests. It should also be clear by now that using wood is not climate-neutral but fuels the greenhouse effect. I'm repeating these points here because forest agencies are currently going down exactly this path. They are recommending to politicians that more conifers be planted. Every tree that is cut down to make roof joists or furniture is supposedly a win for the climate. After all, the carbon dioxide in the wood of these products is stored there forever, right? And that brings us back to the studies carried out by Naudts and her colleagues.

There are a number of downsides to planting conifers. Plantations of spruce, pine, and Douglas-fir are easily damaged when storms blow and when insects attack. In the heat of 2018, many square miles of coniferous forest in Europe, and indeed worldwide, were attacked and killed by beetles. After the wood had been harvested, numerous clear-cuts remained that are now emitting carbon dioxide in the quantities I mentioned earlier.

The pinnacle of misunderstanding and environmental destruction is reached when, for reasons of environmental protection, wood is burned in power plants that used to be fueled by coal. This what the British power company Drax is doing. According to the German website "Plattform-Wald-Klima," Drax imports wood pellets from places such as the southeastern United States. The wood comes from bottomland forests that are clear-cut. The peaty soil then releases massive amounts of carbon dioxide into the air. In 2018, 7 million tons of wood pellets were burned in combustion chambers to produce carbon dioxide and water. According to a British government study analyzed by the organization Carbon Brief, burning wood pellets is up to three times more detrimental to the carbon balance of the planet than burning coal.[94] So why do people do it? Because on paper and according to current rules, wood is still considered climate-neutral no matter what its end use.

In the future, Drax hopes to capture the carbon dioxide emitted from its power plants' smokestacks and—this is difficult to believe—sell it to breweries, which would use it to put the bubbles in beer.[95] Apart from the fact that you could never sell that much beer, as every beer drinker knows, the carbon dioxide in the bubbles escapes either directly from the bottle or, somewhat less directly, from the drinkers—and the carbon dioxide is, once again, back in the atmosphere.

MY OWN PERSONAL goal is that, in the future, we will protect the climate by using less while simultaneously allowing as many forests around the world as possible to revert to their natural state. Primeval forests are our most powerful allies in the fight against climate change.

27

GOOD THINGS TAKE TIME

NINE THOUSAND FIVE hundred and fifty years ago, our ancestors were still living in the Stone Age. Agriculture had just been invented and was not yet widespread, and it would be another five thousand years before famous individuals such as Ötzi the Iceman appeared on the scene.

Yet a seemingly unspectacular event happened that resonates to this day. A single spruce seed fell to the ground in the mountains of Sweden and germinated. Old Tjikko was born. Back then the little tree had no name and was one among millions. Its distinguishing characteristic, as it turned out, was its resilience. Despite many changes in climate, despite catastrophic weather events and hungry animals, Old Tjikko survives to this day and is considered to be the oldest living tree on Earth. It was clear that, tree lover I am, I would one day have to visit this Methuselah.

On May 10, 2018, I finally got my wish. After we drove through Stockholm and Mora, the road got narrower and narrower, the traffic ebbed away, and the distances between those typically Swedish red-and-white houses increased. Finally, I made one last turn into Fulufjället National Park. The narrow park road wound its way past ancient trees and raging streams swollen by rapidly melting snow. Walking season had only just opened, and the parking lot in front of the national park office was empty but for an all-terrain vehicle belonging to Sebastian Kirppu, who was going to be my guide for the day.

Sebastian had called while I was still on the road to ask if I wanted coffee or tea before we set out. Good, I thought, that promises a relaxing day with no need to keep an eye on the time. Munching on cookies, we looked around the visitor center, empty of people because it was still closed for the season. But the outdoors was calling me. I was so excited to see Old Tjikko.

THE SUN WAS shining on the slopes of the fjäll (fell). The day was warm, unseasonably warm. The advance information from Sebastian had led me to expect a strenuous ascent up the mountain with us toiling upward on snowshoes. But, in fact, most of the snow had already melted and the pathways were drying out under temperatures over 68°F (20°F). We ended up with a hike rather than a climb, even if the hike was quite a challenging one.

Numerous narrow boardwalks crossed small patches of heath between stands of trees. We'd walked for no more than thirty seconds along one of the boardwalks when Sebastian turned off and stamped through a couple of patches of snow with me in tow. We stopped by a half-rotted stump where a pine tree had snapped about 6 feet (2 meters) above ground level. "We just had to take a look at this," Sebastian said as he pointed to a small, bright green lichen that looked like a miniature bush clawing its way out of the stump. "Just don't touch it. In the old days people used it to poison wolves." Ignoring his own warning, Sebastian prodded the lichen with his index finger.

When Sebastian told me it was threatened with extinction, I began to feel sad for the little life-form. The habitat requirements for the wolf lichen are so specific that these days, when modern forest practices rule, there is simply no place left for it to grow. This lichen needs one thing above all others: time. First, a pine needs to grow in the forest. (A spruce won't work.) This pine must grow to be ancient, many hundreds of years old at least. When it dies, its trunk must then gradually break down over the course of another few hundred years. It slowly shrinks and develops cracks and crevices, but it doesn't rot because the wood of pines contains so much resin. Now, finally, the wolf lichen can settle in and unfurl its startling green structure. I stared at the stump and a wave of emotion swept over me. Where in my home was there a forest where trees are left to grow this old? There are probably no wolf lichen in Germany, but there would be other life-forms that also need old, timeless forests.

The two of us were quickly deep in conversation about modern forest practices. The clear-cuts not only in Sweden but also national parks in Germany demonstrate the widespread lack of an in-depth understanding of the complexity of such fragile habitats. But there was no time to stand there and mope because, just a short distance farther on, Sebastian made a second stop. Old Tjikko was simply going to have to wait for another hour or two.

Once again, we were examining a stump, another pine, but this time one that had been scorched by fire. Try as I might, I could not see anything special about it. Then Sebastian pulled out a magnifying glass and pointed to tiny black spots—another lichen with an unbelievable need for time. Unlike its green colleague, however, it required an additional ingredient: charred wood. But not just any charred wood. It needed charred wood that was at least a hundred years old and still had old, intact wood underneath it. Here and only here did the lichen feel at home—in a place where it was easily overlooked.

Because most normal visitors (and I count myself among them) are so enchanted by the colorful flowers in the mountain forests of Sweden, they don't even notice these slow-growing lichens. No one will mourn their passing if they disappear when the ancient forests are cut down and replaced by plantations. No one other than a person like Sebastian, of course, and he is doing all he can to keep more forests from being devoured by the increasingly rapacious forest industry. With this goal, he visits ancient forests that are on the chopping block. Usually the foresters involved downplay the issue and deny the existence of rare species in their forests. And if some do live there, in the best-case scenario, they agree to allow small islands of trees to stay standing—spreading no more than 20 feet or so (a few meters) in any direction, they are far too small to sustain living communities.

Sebastian likes to make the comparison with a city if you were to tear down all the high-rises but one. Then, all the residents would have to move into the one building that is left, which, of course, would be impossible. Something similar happens with the thousands of species that suddenly lose their homes. Sebastian sometimes manages to put a stop to the cutting at the last minute, but usually not. And for this reason, he feels frustrated that he is not doing enough, although he really should be very proud of his work.

BEFORE WE MADE our final ascent up the steep slope to the high plateau, we rested for a while by an impressive waterfall. Swollen by snowmelt, the waterfall thundered over the cliff edge, the huge quantities of water creating clouds of mist that sparkled in the sun. Walkers in sneakers make it to this point. Then, after taking a quick selfie and tossing their empty soft drink cans into the landscape, they turn for home. You can see how nature can become a tourist attraction to be ticked off the list. All people are interested in is a photograph to take back to impress their friends. Most visitors are blissfully unaware that rare falcons nest on the cliff face next to the waterfall.

Now the path got steeper and we were alone. Up and up we went over fields of snow and scree to the high plateau. From there, we had a breathtaking view for miles over the national park. Its boundaries were, unfortunately, only too easy to see. Clear-cuts showed where commercial interests had destroyed the green canopy of nature and left a scarred landscape. "It's over there!" Sebastian pointed to a small green triangle on the horizon. We trudged over to our goal, dry lichen scrunching under our boots.

I had finally arrived. In front of us stood a windswept spruce that rose up out of a cushion of green twigs. The landscape around it was strewn with boulders, which underscored the barrenness of this high plateau. What went through my head? Even though no tears came, I was still extremely moved. For a moment I was speechless as I thought about how long this tiny scrawny tree had held out up here. Almost ten thousand years had passed since it germinated from its seed. Mammoths had died out, Stonehenge had been erected, and the pyramids had been built. The climate had fluctuated from cold to warm and back again multiple times, but, unaffected by any of this, the spruce was still standing intact today in the place where it had been born. Other than the changes in climate, it had certainly not been aware of any of these things, as they belong to the human experience.

Old Tjikko could grow to be as old as it is only because it grows especially slowly. Tjikko's rate of growth is dictated by its environment. The growing season up here is extremely short. Winters are long and hard. There is not much time to get even a small amount of photosynthesis going. Over and over again, heavy loads of snow bent the tree's tiny trunk, so a side branch took on the job of growing upward, creating a new vertical trunk. This means that the "tree" we see today is only a few hundred years old. The true old spruce is to be found in the roots and in the brushy growth covering the ground.

As I looked at the tree, I once again asked myself what really makes a tree a tree. Is it the trunk, which we usually think of as being the most important part? Or is it the roots, which have survived for thousands of years and are where the old spruce has probably stored all that it remembers? At the moment, I'm tending toward the latter.

AS SEBASTIAN AND I ate our packed lunch—"polar bread," a sort of soft crispbread, accompanied by blueberry juice and cheese—he told me the park authorities were considering creating a marked trail to Old Tjikko. Many tourists came to the park to look for the tree in the mountains and went to the visitor center, angry and disappointed, to complain when they couldn't find it. Old Tjikko was, after all, the reason for their visit.

A marked trail? I did not like this idea at all. As I looked at the fragile spruce, I conjured up images of thousands of souvenir seekers wanting to take home a twig as a trophy in addition to the obligatory selfie. And that wouldn't turn out well in the long run.

Up until now, only the tree's general location has been marked on the map, and there's a thin white rope on 1-foot-tall (30-centimeter) stakes keeping people at a distance of 16 feet (5 meters). This "barrier" is supposed to keep people from walking all over the tree's delicate roots. Outside the rope perimeter, a number of lichens had already been trampled into the boggy ground. As Old Tjikko's roots certainly spread underground at least twice as far as the trunk is high, the tree is already being damaged by visitors. And by me.

Then I began to get concerned. With my size 14 (European size 48) feet, I, too, had dispatched a number of delicate lichens into nirvana. I had an even more worrying thought. If I made more people aware of this treasure, wouldn't I be guilty of encouraging increased crowds of tourists in the future? I found this idea very troubling. Would it be better if I stopped reporting on nature's jewels, on intact ecosystems that give hope and pave the way for many people to reconnect with nature?

Perhaps there were better ways to do this?

Right now, there are is a guided walk to Old Tjikko every day in the hiking season. Couldn't it be left at that if a waiting list was set up? Pop concerts get sold out, after all. Why should it be any different when the star attractions are trees? Another alternative would be to deny access to everyone. I don't see that as an option, however. If you protect nature by keeping people out, they will lose interest in preserving ecosystems.

BACK IN THE valley, we made another stop in the national park offices. Helena, Sebastian's girlfriend, went in and came back out again with a piece of sausage in her hand. "That's for the Siberian jay," she said. In German, this bird is called "the jay of hard times" because when the weather is especially cold, Siberian jays have to fly as far south as Central Europe if they are to find anything to eat. For centuries, their appearance has signaled extremely harsh weather conditions such as freezing temperatures or huge amounts of snow. That certainly meant hard times for poor people struggling to survive in rural areas, and that is how the bird got its name.

In Fulufjället National Park, however, the birds cheer the tourists up. If scrawny lichens are not sufficiently interesting on a day out, then the tame birds can bring some joy to disappointed hikers—especially those who haven't been able to find Old Tjikko because the information on the map is so vague.

And this brings us to the heart of the problem here. Many of the visitors are not drawn to and fascinated by the wonderful landscape with its breathtaking views, and they have little interest in the abundance of highly specialized life-forms. What attracts them here is the great age of the scrawny little storm-tossed tree that ekes out its existence high up on a mountain ridge. Visually, the spruce is nothing special; it's all about its history. Its attraction lies in simply knowing that it has been fighting for its life for 9,550 years and may possibly survive for a few thousand more. Small spruce trees are, after all, a dime a dozen. It is only its extreme old age that sets Old Tjikko apart from the other elderly spruce in the area.

MY TOUR WITH Sebastian, however, was not yet over. He was keen to show me more ancient forests close by. We did see them, but what we saw most was countless piles of logs from old-growth trees—stacked in front of enormous clear-cuts that had completely destroyed these unique slow-growing ecosystems. The worst, according to Sebastian, was that this wood came with a seal from the Forest Stewardship Council (FSC), the mark of especially environmentally friendly and sustainable forest practices.

I use this seal myself in the forest I manage in Wershofen. Given what I have observed in recent years (the FSC also certifies wood from clear-cuts in the Eifel National Park near me), I am no longer sure if the label means anything. Can there be anything worse than wood from clear-cut ancient forests? If the FSC's seal not only does not stop clear-cutting but continues to certify the wood even though it's obvious what's happening, then isn't it time to look for an alternative? But what would that be? There is the seal from the Programme for the Endorsement of Forest Certification (PEFC), which has even lower standards than the FSC. Then there is a glaring gap. All you have left is timber harvested with no external oversight by non-governmental organizations.

We drove home with mixed feelings. However, we did have one last positive conversation during our farewell meal in a small restaurant tucked away in the wilds of the Swedish countryside. All over the world, activists like Sebastian, who fight alone, achieve much, and yet at the end of the day are still frustrated. Wouldn't it be great if this group met once a year simply to exchange ideas? Without an agenda, without a specific goal, simply in order not to feel so isolated? That evening we agreed to hold just such a meeting, and I am already excited for all the lone wolves—let's gather for a group howl!

28

IN SEARCH OF BOTH THE FUTURE AND THE PAST

An ancient tree like Old Tjikko is an unparalleled demonstration of the most important issue in nature. The sheer, endless amount of time needed to fully develop an exquisitely balanced ecosystem. If a single tree can grow to such a great age, what about an old-growth forest? How old does a collection of trees need to be to earn this designation?

This question is important because, in Germany at least, we desperately need to convert plantations into old-growth forests. Plantations simply do not offer the species diversity, climate control, and opportunities for rest and relaxation found in old forests. In addition, we feel responsible for correcting the excesses of recent decades. But to do this, we first need to have the answer to this question: what does an old-growth forest look like?

This is the idea I wanted to bring to life for people, and my books have given me an opportunity to do so. Actually, that's not strictly true. It was readers and the overwhelming interest they showed in *The Hidden Life of Trees* that made that and my later books into international bestsellers.

The books were read in Canada, as well, and a call for help from that country reached me via email. It was from Frank Voelker, the band administrator for the Kwiakah First Nation in British Columbia. The band has only twenty members, which makes it difficult for them to stand up to the timber industry. A big victory for the environmental movement—the long-sought protection of the Great Bear Rainforest on the northern coast of the province—is, indirectly, the source of their misery.

THE GREAT BEAR RAINFOREST is the largest intact temperate rainforest in the world and for now, it seems, it is safe from the clutches of the timber industry. For twenty years, the indigenous population and conservation groups fought not only for the trees but also for the animals. Grizzly bears were a prime target for trophy hunters who wanted to upgrade their living rooms by adding hides and skulls. The Great Bear Rainforest Foundation bought up all the hunting licenses and let them lapse so that, at least within the boundaries of the rainforest, the bears were no longer in the line of fire.

Since 2017, no hunting licenses have been issued to trophy hunters for grizzly bears anywhere in British Columbia. One year before that, 85 percent of the Great Bear Rainforest was protected in perpetuity. Since then, nearly 12,000 square miles (30,000 square kilometers) have been closed to commercial interests—even dams for hydroelectric projects are forbidden. At the time, I was very happy to hear this heartening news, which is why I was so concerned when I received the email from the Kwiakah First Nation. The forest and timber industry was now looking for alternatives, and they were clearly finding them by increasing logging in available areas to the south.

Although they were not exactly finding them. They were being granted them by the provincial government to compensate them for what they had lost when the new nature preserve was established. One of the forested areas affected by a massive increase in timber harvesting is Phillips Arm. It is part of the traditional territory of the Kwiakah and covers nearly 200 square miles (about 500 square kilometers) in an area where the original inhabitants are still fighting for their rights. The band is amenable to a compromise. They do not wish to forbid any kind of commercial forestry, but they would like to see more environmentally friendly forest practices. The Kwiakah, too, profit from grizzly bears—but only when they are alive. Bear-viewing is part of the low-impact tourism opportunities offered by the band.

Where forests are clear-cut, heavy rainfall washes the exposed soil into the nearest stream. Salmon cannot survive in sediment-filled streams and the waterways become devoid of life. Now, the grizzly bears can no longer find salmon to feast on in the fall, salmon they need so they can lay on a thick layer of fat to survive their deep winter sleep. And so the grizzly bear population declines and tourists miss out on a trip of a lifetime. But more than that, the whole food chain—from insects to small rodents to bald eagles—falls apart. All that in an area where the residents, and the government, depend heavily on paying guests.

FRANK SUGGESTED THAT I come and visit the band to support its cause. We wanted to tour the different forests—untouched old-growth, selectively logged areas, and clear-cuts. The goal was to see how logging might happen in a more environmentally friendly manner in the future and, most importantly, to see if we could influence the forest policies of the provincial government. I was happy to agree to a trip in October and was very excited to spend time in an authentic old-growth forest.

Frank organized everything. The very first thing was a greeting ceremony the evening I arrived in Campbell River on Vancouver Island, where the band's headquarters is currently located. The next morning, Frank drove us, bright and early at 6:50 a.m., to the marina. A skipper was waiting to take us out to the reserve in his aluminum boat. The crew turned out to be fellow travelers: two journalists, three foresters, and the band chief. Decked out in his rain jacket and wool cap, Chief Steven Dick did not look anything like what I was expecting. But my expectations had been influenced—and not in a good way—by youthful readings of the German author Karl May with his fanciful representations of Native American heroes. Also, Chief Steven seemed a bit shy at first, although this initial impression was quickly dispelled by his extremely likeable demeanor and understated sense of humor.

The boat ride was quite rough. It was raining and a cold wind was whipping up waves so we were constantly hitting a wall of water as the boat crashed its way from one peak to the next. The windows in the cabin were fogged up inside and out, and we could make out very little of the gray landscape as we motored past. Welcome to the Pacific Northwest rainforest. Where I live in the Eifel, we have about 30 inches (76 centimeters) of rainfall a year. Here the total is over 260 inches (660 centimeters). (That would be 800 liters of rain per square meter a year in Germany, as opposed to 4,000 liters per square meter per year in the Great Bear Rainforest.)

WHEN THE BOAT turned into a bay after a seventy-five-minute crossing, we could make out our living quarters by a small marina: the Sonora Resort. It was nestled into a mountainside and was built almost entirely of logs. A welcoming party was there on the dock. Two employees took charge of our luggage while we went to check in. Because we were special guests, we were given a complimentary drink as we waited for the manager. We were shown around the resort, including the water-treatment system, and then it was time for us to get back into the boat. The purpose of our visit, after all, was to see the original territory of the Kwiakah, and that was another 12 miles (20 kilometers) or so away on the mainland.

As we arrived in Phillips Arm, the inlet revealed itself in brilliant sunshine. The rainclouds were not forecast to return for the next two days and so we had picture-perfect weather—although not a picture-perfect landscape. The wounds inflicted by the timber industry could be seen on every mountain slope. Logging roads zigzagged down the mountainsides and the forest was divided up into parcels of different-aged trees. Only along the shoreline could you still see the pitiful remains of what had once been old-growth forest.

I was familiar with such scenes from Europe, where whole forests are managed by dividing them up into sections so you can clearly see the mosaic even from a long way off (you can also see this using Google Earth or a similar program). Even though I knew that the timber industry in Canada had been recklessly clear-cutting for decades, I had hoped, at least in such isolated areas as Phillips Arm, to find some remnants of the original forests. Chief Munmuntle (Steven's Kwiakah name) fueled this hope. He welcomed us to the land of his ancestors with a short speech. Frank was surprised because this was the first time he'd heard the chief make a speech like this. Chief Munmuntle made mention of the old-growth forests that we wished to view. I'll jump ahead here: unfortunately, we found no old-growth forests during the two days of our visit. Only narrow strips no more than 165 feet (50 meters) wide had been left more or less untouched around the edges of the clear-cuts, the few lonely ancient trees that had survived.

Two representatives of big timber companies, Tanja and Domenico, were brave enough to discuss the forest situation with us. This is how they described what was happening on Canada's West Coast. The province, which owns most of the land, needs income. It generates this through the sale of timber licenses, which are auctioned off to the highest bidder. The licenses apply to defined areas that cover hundreds, sometimes thousands, of square miles (sometimes up to 3,800 square miles/10,000 square kilometers) from which the companies must harvest a certain amount of timber within five years. Must? Yes, because the authorities cash in later; that is to say, after the trees have been cut. That's when a stumpage fee falls due. It is charged per cubic meter of wood and the amount ranges from US$14 to $39 per cubic meter at the current exchange rate, depending on the quality of the wood and the state of the market. If a timber company buys a license but then cuts too little wood—or no wood at all—they lose their license.

But no system is so terrible that it cannot be presented in glowing terms with good publicity. In the case of Canada's West Coast, this is the spiel: a timber company that logs an area only once every eighty years and then leaves it completely undisturbed has fewer impacts and is far less disruptive to sensitive species such as the grizzly bear. Moreover, clear-cuts follow aesthetic guidelines that require very little of the clear-cut to be visible from the shoreline. The reason—you've guessed it already—is tourism. Visitors should still be able to experience the idyllic landscape of British Columbia, the one we all know from television.

From what I saw, however, that's not working very well. As we traveled in the boat across the ocean to the band's ancestral lands, we could see young forests everywhere, broken here and there by brown patches—fresh clear-cuts. Along the shorelines, there were narrow strips of older trees that couldn't hide the environmental damage behind them. The sea lions dozing on the rocks along the shore were not enough to make up for the devastation.

Later, when we were walking through the forest, I asked the two foresters, Tanja and Domenico, whether less-invasive selective thinning might not be better. Certainly, less invasive is always better, but I had to consider that the forests would then be disturbed much more often. Moreover, an area ten times as large would need to be harvested to get the same amount of wood, because when a forest is selectively logged, no more than 10 percent of the trees are harvested. And of utmost importance: the network of logging roads would have to be maintained regularly and even be expanded. People like me who want to log using horses needed to have some way of getting to the forest.

This argument is not wrong. In Germany, for instance, for every square kilometer of forest, 13 kilometers of logging roads are built that cut the ecosystem up into completely separate pieces (that is equivalent to 20 miles of roads for every square mile of forest). Some animals, especially ground beetles that are particularly sensitive to light, do not cross these treeless cuts through the forest because it is too bright for them on logging roads. Moreover, the compacted roads interrupt the flow of water underground, which leads to waterlogged areas on the slopes above them and dry areas below. The question is, could we, in Germany, manage with fewer logging roads? The answer is a resounding yes.

I also think the more regular disturbance that results from selective logging is tolerable. Animals adapt to such practices, as the extreme example of military training grounds demonstrate. You get the best wildlife viewing opportunities when the tanks are having target practice because the deer in the area well know that as long as the tanks are firing, there are no hunters around to take their lives. People are disruptive only when they are perceived as predators, as national parks in Africa and North America demonstrate in quite an impressive fashion. In Yellowstone National Park, for instance, bison allow tourists to approach within a few feet because they regard people as just another harmless species that calls the grasslands home. The comparable situation in British Columbia would be that the appearance of loggers would not stress the animals if hunting were banned whenever loggers were around. That should be a fairly obvious solution in an area where wildlife tourism is important.

The Kwiakah now have two options. Either the province and the timber companies join with them voluntarily to find a less destructive way to log the area or the band fights for its rights in court (which would cost many millions of dollars, money the band does not have). And so, for the moment, the first option is the only one open to them.

THE TASK FOR the future, therefore, must be this: How can commercial interests transition to less-invasive thinning practices in these forests, and, additionally, how can we set aside sufficiently large areas protected from any kind of industrial forestry? The solution seems so simple. The land still belongs to the original inhabitants, the First Nations that live along the coast. They have a completely different understanding of the forest than many of the newcomers. They use wood to build their homes, for instance, but they often don't fell whole trees. Instead they cut individual planks from the trunk so they don't kill the tree.[96] That is invasive, to be sure, because it severely damages the tree. However, this way the tree remains alive and the sensitive structure of the forest is left mostly intact. There are trees standing today that have been mapped and protected as cultural sites—a testament to how little impact this practice has on the lifespan of western red cedars in coastal forests.

In contrast, western red cedars are so coveted by the timber industry that it's almost impossible to find mighty trees today. They are cut down even on steep slopes and then, because there are no roads, flown by helicopter to the nearest inlet (where they continue their journey, rafted up in log booms). One of the consequences is that many First Nations people can no longer build traditional canoes because there are simply no trees left of a suitable size.

FOREST MANAGEMENT THROUGH the eyes of First Nations should lead to a picture we recognize from selection cutting or continuous cover forestry in Europe (forests managed this way are called *Plenterwälder* in German). In these forests, commercial forestry happens in and around stands of old-growth trees, tree families are kept together, and individual trees are allowed to grow very old. Here and there, a sturdy mature tree is felled, but otherwise the forest is left untouched. A forest like that, interspersed with protected areas where no logging is allowed, would help restore the forests around Phillips Arm to what they once were.

In Germany, one example of this type of forestry can be found in the Lübeck municipal forest, where the indigenous population (that is to say, all of us) is restoring a hint of the original forest. Admittedly, this process takes time—forests need hundreds of years to recover from the disruption and destruction wrought by people. At what point is it even possible to talk of an authentic old-growth forest? Think of those interminably slow lichens around Old Tjikko that take centuries to appear on the remains of old trees. How many other, similarly slow life-forms are out there? Given our current level of knowledge, can we even begin to suggest a time frame?

I think we can, and I have a practical suggestion. At least one generation of trees must have been left to grow without any human interference. Depending on the tree, this could take as long as five hundred years. Does that sound like a long time? Consider that the first generation would have grown up in the era of plantations and clear-cutting, and will therefore have grown up in unnatural conditions. The slow growth I have described would start in the second generation, which would be growing up in eternal twilight under the protective canopy of the mother trees. The full process of creating an old-growth forest with its ancient diversity of species could be expected to start at this stage at the earliest.

MEANWHILE, WHAT IS happening in the Kwiakah's forest are cosmetic corrections that affect only the aesthetics of the forest. On the boat trip back, one of the foresters pointed to the slope on the opposite shore. "You can only see 15 percent of the clear-cut," he said proudly. The rest was so cleverly done that it disappeared behind the hills in front of it. What tourists see is nothing more than a green backdrop that imitates old growth.

When I arrived home, I got another email from Frank. First, he thanked me because my visit had given the band hope for the future, but then came the bad news. A timber company representative who had met with us had informed the Kwiakah that a portion of their forest would soon be fertilized by helicopter as a pilot project. Frank felt this was a slap in the face, especially as the representative had not mentioned this when we were in the forest together just a few days earlier.

Fertilizer sounds harmless enough. Don't we fertilize agricultural fields and our roses? In cultivated landscapes that might be fine if it's not overdone. It might even be necessary. In the forest, however, it's a catastrophe. Trees don't like to grow quickly, and they can grow old only if they have spent at least a century as young adults under their parents' protection. The spruce, firs, and cedars planted in the forest industry's clear-cuts are already growing rapidly out in the bright sun. When the whole thing is now shifted into even higher gear by the application of fertilizer, there's little difference between these trees and pigs raised in factory farms—racing through life and ready for slaughter in no time at all.

The rest of the ecosystem is even more severely damaged, because all parts of the forest are subjected to the same treatment. In areas that have naturally low nutrient levels, sometimes with highly acid soils, highly specialized life-forms are completely wiped out by this rain of dust from the sky. With these programs, British Columbia is reducing its northern rainforests to nothing more than fields of trees awaiting harvest.

I decided that my visit would not be a one-off event but the jumping-off point for a program of long-term support for the Kwiakah. Nowhere else on my journeys had I felt such strong ties between people and nature than I did when I was with the band. I wanted to help keep it that way.

29

PROBLEMS HIGHLIGHTED BY BIAŁOWIEŻA

A GROUP FROM POLAND attracted international attention recently because they were campaigning for a particularly exceptional piece of nature: the ancient forest of Białowieża. The debate that rages around Białowieża sheds light on the plight of many forests worldwide. The forest lies on the border between Poland and Belarus in a region where the climate is harsh, too harsh for beeches. Although beeches shaped the original forests of Central Europe, you don't find them here because the winter is much too cold and lasts much too long. The trees that dominate in Białowieża are oaks, lindens, hornbeams, maples, and spruce.

The Polish government, or rather the ruling party, Law and Justice (PiS), clearly thought it wasn't worth protecting nature in Białowieża, and they allowed extensive logging in the forest around the national park. When protests erupted, the other side trotted out an old argument often used by foresters: bark beetles were going to eat up the forest and the only way to avoid this catastrophe was to remove infested trees (and then, naturally, sell the wood). The argument had some basis in reality. In recent years, an enormous army of engraver beetles, tiny members of the bark beetle family, had indeed gathered and was destroying bark and eventually whole trees in large stands of spruce.

The engraver beetle is named for the symmetrical galleries it and its larvae excavate under the bark. Engraver beetles love spruce, or more precisely, they love the growth layer between the tree's bark and its wood. It is juicy and packed with nutrients, and people can eat it, too. And here is the problem. Healthy spruce can defend themselves by drowning each attacker in a drop of resin as soon as it drills into the bark. In hot, dry summers, however, which are becoming increasingly common thanks to climate change, the trees weaken and release scents that signal they are stressed. The beetles smell these distress calls and fly on over to attack and kill the weakened spruce. Then they move on to the next available tree and often end up making a meal of healthy trees, as well. The sheer number of attackers means that the tree cannot defend itself from each beetle drilling into it, and it gives up. And so bark beetles can destroy large spruce monocultures. However, while this is happening, pathogens are spreading through the beetle population and eventually disease wipes out the beetle army.

So, was cutting down and removing trees a reasonable response in an area that includes one of Europe's last remaining old-growth forests? Or, as many foresters argued, perhaps it wasn't an old-growth forest, after all.

These are tough questions that even give people who want to protect forests pause. Here are the facts: A small section of the ancient forest, about 40 square miles (100 square kilometers), was declared a national park in 1932. A larger section of about 400 square miles (1,000 square kilometers), across the border in Belarus, was protected in the same way in 1991. The whole forest on both sides of the border was recognized by UNESCO as a World Heritage Site, joining the exclusive company of jewels such as the Great Barrier Reef in Australia and Yellowstone National Park in the United States. In addition to the ancient trees, there are more than twenty thousand species that call this forest home, including the mighty wisent, which was close to extinction and remains critically endangered.

Let's stay on the Polish side of this conservation area. The ancient forest spreads far beyond the arbitrary borders of the national park. In total, it covers more than 230 square miles (600 square kilometers), and the rare species it contains make it so valuable that it has been designated a Natura 2000 site by the European Union, which affords it protection so that only limited intervention is allowed. This official recognition is intended to protect the last at least partially intact ecosystem in Europe. The authorities accept that really large national parks like the ones in the United States are not possible in Europe, and protected areas like this are a compromise because they allow some use of the landscape—as long as it is not too detrimental to the forest.

But detrimental use of the forest is exactly what happened in Poland, and the intent to inflict the damage was made public. The bark beetle was used as an excuse, but the real reasons were clear. Cutting trees would satisfy two needs.

First, those responsible did not want to leave so much timber in nature's hands and, second, they wanted to show the European Union that they didn't give a hoot about restrictive rules aimed at protecting the environment. In 2016, with one stroke of his pen, Jan Szyszko, who was the minister for the environment in Poland at the time, tripled the area available for logging and gave the forest industry until 2023 to take almost 260,000 cubic yards (200,000 cubic meters) of wood from the protected area.

Despite intense coverage by the international press, heavy machinery began to cut down hundreds of trees a day. The operation was billed as a rescue mission, and cutting down trees was presented as the only option to destroy the bark beetles that were supposedly threatening the rest of the forest.

In Białowieża, however, there are many reasons to stand back and watch natural forces play out their roles without intervening. The forest here is not a spruce monoculture but an old-growth forest where many different species of trees grow mixed together. If the spruce are killed by bark beetles, all that means is that the forest will be thinned without creating any open areas devoid of trees. Moreover, there is the forest's protected status. It was put under protection to ensure the continued unrestricted interplay of natural forces, which includes, and must include, some that are considered undesirable from a human perspective. Besides that, nature has managed such processes successfully for millions of years—it does not need us.

The huge clear-cuts, therefore, carried out under the guise of rescuing the forest, did exactly what people feared the beetles would do: they ensured the wide-scale destruction of a protected forest. There was no question. I needed to go and lend my support to the local activists.

PIOTR AND ADAM, researchers and environmentalists, picked me up from Warsaw airport in a van and we drove for hours toward the border between Poland and Belarus. The vehicle almost flew over the potholed roads; it was only because I was having such an interesting conversation with the two of them that I could forget my worries about arriving safely at my destination. We arrived at Białowieża late in the evening, and my first stop was not the forest but the protest camp.

Contrary to my expectations, we were not greeted with colorful waving flags nor did we find a tent encampment. Instead, the protectors of the forest had set themselves up in a big old house. If you want to hold out for years that is perhaps a better choice. We were greeted warmly and invited to take a seat at a large wooden table with benches on either side. Here, we had coffee and cake before our real mission started: offering support merely by being there. Of course, we didn't just sit facing each other in silence but talked with everyone about the problems and successes they'd had.

A camera team from a local television station was also in attendance. I had been warned that they were not totally on the side of the protesters and that their views were shared by many of the residents near the protected area, who, like the Canadian foresters I had met, made their living from the forest and the timber it provided.

We spent two nights at Wejmutka Manor, a comfortable wooden guest house on the outskirts of Białowieża village not far from the national park. The owner supported the protest movement, and so it was no surprise that, on the second evening of my stay, she hosted a conference attended by scientists, conservationists, and friends of the national park.

I HAVE TO admit that I couldn't keep my eyes open, not because the conference was boring, but because we had left early that morning to see wild wisents. If we were in Białowieża, there was no way we were going to miss seeing these impressive animals. And that meant getting up at 3:30 a.m. to have the best chance of seeing them as soon as it began to get light outside.

Piotr, Adam, and I stood in front of the hotel still blurry with sleep, waiting for the park ranger. He drove up in an old Škoda and got out, outfitted from head to toe in military-style camouflage clothing. He murmured a brief greeting and got back in. We jumped into our car so as not to miss our appointment with the wisents. We stopped at the edge of a small village. It had begun to get light and the landscape was wrapped in fog. The grass was soaking wet and it wasn't long before our shoes and socks were, too. We walked quietly and carefully along a path through the grassy landscape, following the ranger. "There." He pointed to a bank of fog. At first, we saw nothing, but then, gradually, three powerful bodies took shape. Wisents. We excitedly took turns at the telescope the ranger had set up for us. Three shadows in the fog, the horns clearly identifiable—then they turned around and the fog swallowed them once again.

"Sometimes we don't see any animals at all," was all the ranger said, making it clear to us that we had been really lucky. When I asked him what the animals ate, he said that naturally they ate mostly grass. To keep the local farmers on the side of the wisents, the government gave them grants to manage their pastures especially for these wild cattle. Wisents prefer to eat old varieties of grass that contain fewer nutrients than new varieties, which are then made even more calorie-rich with the application of fertilizer.

I'm talking about grass here, but aren't wisents forest animals? And right there is one of the problems presented by fragmented modern landscapes.

ORIGINALLY, THERE WERE indeed grass-rich landscapes in Central Europe, along the banks of rivers. In the days before climate change, you could watch as ice floes floating down rivers swollen by snowmelt destroyed the trees trying to grow in these meadows, which created these open spaces. Apart from the meadows, there was grass for large herbivores up above the tree line in the mountains and around the margins of marshland. That was it.

We, however, don't want to give these areas up and so we force animals like the wisent into protected forest areas, where they are not equipped to survive. And that is how we end up with the agricultural production of grass, and in winter hay, to keep them alive. The wisent and the farmers are happy, but these ancient cattle are no longer living wild and free as they once used to do. What our little travel party experienced, therefore, wasn't much more than an excursion into a big safari park. To change that, we would need much larger protected areas. But the Polish government was in the process of logging the forest around the park, and that was not only affecting the wisents but also many other animals whose tracks we also saw on numerous occasions, such as lynx and wolves.

AFTER OUR MORNING safari, we returned to the house where the activists were staying and from there set out along a logging road for our hike through the forest. We passed mighty, five-hundred-year-old oaks. As the first mosquitoes began to buzz around us, we turned off and walked right into the forest. After a short distance, wide tire tracks appeared, some of them sinking down almost 3 feet (1 meter). They led off deep into the forest, and their trail was littered with huge fresh tree stumps. Clearly, not long ago, a big machine had carried off felled trees—from the protected forest. These massive clearings left by the loggers were everywhere. The ground was sprouting fresh green growth, a sign that an extraordinary amount of light was reaching the forest floor. We took photographs for the press and for my social media account.

Along the next logging road, we found dead giants neatly stacked, just waiting to be carried away. The protectors of the forest had counted the rings on many of them to calculate their age, which they had then spray-painted on the downed trunks. That was how they documented that very old trees were being illegally logged and sold. We took a photograph for a banner to promote the conservation of forests and returned to the village of Białowieża.

HERE'S WHAT I learned from my journey to Białowieża. The forest is very old and worthy of protection, but it is not a true old-growth forest, at least many parts of it are not. It's easy to see signs of the reforestation programs of previous decades. And, even if large numbers of oaks and other deciduous trees have not been exploited for more than a century, another four hundred years or so probably needs to pass before Białowieża can truly be considered an old-growth forest. Only then, according the description I gave earlier, will genuine nature stand here once again.

The need for protection, however, continues. Although this forest has been returning to nature for decades, industrial logging could erase all the progress it has made. Moreover, we have hardly any comparable areas in Europe, areas where no active interventions have taken place for this length of time. I wish we had a forest like this in Germany, but the closest ones we have continue to be critically endangered. The culprit, in the case of the forest I describe in the next chapter, is the production of energy.

30

HAMBI IS HERE TO STAY

THERE WAS REASON to take a closer look at what was happening even on my own doorstep, in this case at a lignite mine near Garzweiler on the outskirts of Cologne. The mine made headlines in 2018. Enormous excavators from the German electric utilities company Rheinisch-Westfälisches Elektrizitätswerk (RWE) have been eating their way through the landscape for many decades, destroying not only complete villages but also the forest, and leaving behind giant open pits that are too large to fill. The plan is to turn them into a giant 8-square-mile (20-square-kilometer) lake up to 650 feet (200 meters) deep. It will take until 2045 to fill the lake with cleaned water diverted from the Rhine.

Water is a big issue. Numerous pumps work to suck out the considerable quantities of groundwater that seep in from under forest and fields from more than a third of a mile (half a kilometer) away. The pumps are needed because the giant rotary excavators have eaten 1,300 feet (400 meters) down into the sandy ground. Groundwater flows into the pit from the surrounding countryside, even from as far away as the Eifel, where I've noticed the effects as the wetland areas in my forest are quickly drying out. The forest is suffering as the fossilized forest below it is being mined. That's exactly what lignite is, after all—fossilized trees.

THE HAMBACH FOREST, an old-growth deciduous forest, has sparked particularly active protests. It could be compared to a valiant warrior taking their last stand. It's all that remains of a forest that once stood on the land that is now being mined, and it is of particular ecological value. Ancient beeches and oaks provide habitat for rare insects and bats such as Bechstein's bat, whose females gather in hollows in ancient trees to raise their young in the summer.

By fall 2018, from what was once 15 square miles (41 square kilometers) of forest, only three-quarters of a square mile (2 square kilometers) remained. The rest had been cleared for more mining. Was a big push to save the forest this late in the game still worth the effort? After all, activists had been doing amazing work for years, leading walks to explain what was going on and building shelters in trees at strategically important points. None of this had done much to help the forest; however, the conservationists did manage to attract attention all around Germany. And this was precisely why that tiny remnant of forest was so important: it was a symbol that highlighted the environmental policies of the federal government.

Meanwhile, the summer of 2018 had highlighted some big problems in German forests. In the dry heat, a number of coniferous forests were going up in flames, including one near Treuenbrietzen in the state of Brandenburg in eastern Germany. Fires had broken out in many different locations at the same time, clear evidence of arson. Despite the massive effort put into fighting the fires, they burned through more than 1 square mile (3 square kilometers) of pine forest. Forest? The media, supported by conservation groups, began to question more pointedly whether the pine forests ubiquitous to the region really were forests. Weren't the trees falling victim to fires actually fast-growing, non-native providers of wood set out in plantations? Pines are like gas tanks, filled with highly flammable substances such as terpenes and resin. In a dry summer, all it takes is a single match and the hydrocarbons they contain ignite in an almost explosive fashion.

The original native deciduous forests of Germany, in contrast, are almost impossible to set on fire, which means that forest fires are not part of the natural ecosystems of Central Europe. The burning plantations around Treuenbrietzen could therefore be seen as torches illuminating the failed management practices of foresters. Representatives of this green guild, however, are not given to self-examination, preferring, like farmers, to demand federal subsidies so they can plant the same trees all over again.

AND THIS IS where the Hambach Forest came into play. It highlighted the sorry state of government policy. While the pines were burning in the east and people there were demanding assistance, to the west, columns of foresters were standing, chain saws at the ready, waiting for the first of October, the day on which the clear-cutting of the forest could begin. No forest agencies were calling for help, organizing protests, or educating themselves about what was going on. And yet, lignite is a particularly dirty fuel, one of the drivers of climate change and summer heat waves, and native deciduous forests are much more stable than pine plantations in the face of rising temperatures.

It was mostly young people who were building tree houses in the crowns of ancient oaks and beeches to protect them from being cut down. Small, airy settlements with names like Oaktown, Beachtown, or Lorién filled the old forest year-round with cheerful life. At least until the state government of North Rhine-Westphalia decided to clear the wood. The official arguments were flimsy. Fire safety regulations gave the authorities no choice. They had to step in.

Clearly the makeshift shelters didn't conform to the building code, but the idea they were a fire hazard was ridiculous. Unlike coniferous forests, deciduous forests do not catch fire when lightning strikes or when the sun shines off old glass bottles. Try to set a green beech twig on fire sometime—you can't. No matter, the minister of the interior, Herbert Reul, ordered the evacuation of the alternative dwellings and set into motion the largest use of police in the history of the state. New roads were cut through the forest and existing trails were widened and reinforced. Bulldozers equipped with hydraulic lifts rolled down toward the tree-house settlements. With the police to protect them, workers used the lifts to get up into the tree houses, leaving behind ground flattened by the heavy machinery and all kinds of garbage. Then they started arresting the protesters, who refused to leave the forest.

It must have seemed doubly duplicitous to the public that as events in the forest were unfolding, the coal commission was assembling a panel whose wide-ranging composition was intended to reflect the social makeup of modern German society. The panel was tasked with investigating how Germany could phase out coal-fired power plants with broad consent from the community. At first, the Hambach Forest did not have a role to play here, although it would later. RWE, however, didn't want to wait, not even a few months. The Hambach Forest was to fall, and it was to fall right away. The company insisted that the trees needed to be cut down even though a variety of experts testified that the current area of the mine would provide enough lignite for years, which meant there was no need to remove any more trees.

I'D BEEN POSTING supportive messages on my social media page, but the time had come to visit the site in person. The trigger was a call from Greenpeace asking if we might organize something together in an effort to get a last-minute stay on the tree removal. In Germany, so many forests are managed and planted that they are no longer natural associations of trees, therefore the conservationists working to protect the Hambach Forest preferred to call it a woodland to make it clear that it was a natural space worthy of being saved. Representatives of Greenpeace and I agreed to join one of the tours Michael Zobel and his wife had been leading for years. Michael is a nature guide, and in 2014 he decided he would offer tours of this endangered woodland. Since then, he has led a tour once a month for people who want to learn more about the Hambach Forest. When things heated up and the last three-quarters of a square mile (2 square kilometers) stood under constant threat of annihilation, he increased the frequency of his tours to once a week, always on a Sunday.

September 30, 2018, was one such Sunday. Together with my family and workers from my Forest Academy, I set off through the dusty landscape of the mining area. One hour before the official start of the walk, hundreds of people were already waiting. In bursts, as the suburban trains arrived in the neighboring towns, the number rose, until finally the crowd numbered more than ten thousand. My part was interviews with the media and a short speech from a car equipped with a loudspeaker. Then, I joined the huge throng walking to the wood, peacefully waving flags and every once in a while chanting, "Hambi stays." We were accompanied by a great number of police officers, who secured the woodland and observed the march from helicopters flying above us. Other than that, nothing spectacular happened.

Although that's not quite true. I was personally overwhelmed by the experience. After weeks of excess by the radical right in cities such as Chemnitz, where foreigners were being harassed and people were being chased, the Hambach Forest was another world. Here the protest culture of the early 1980s that had once been directed against atomic warheads and nuclear power stations was alive once again.

Reporters kept asking me if the woodland really had any chance of surviving. After all, it stood right at the edge of the mine, next to a huge crater from which all the groundwater was constantly being pumped away. And "Hambi" now covered just three-quarters of a square mile (2 square kilometers), an area far too small for the trees to create a proper moist forest microclimate. The woodland was so damaged that its future really did seem to be questionable.

The answer was simple because the dry hot summer of 2018 functioned as an extreme stress test for the trees.

During the October demonstration, it was clear that the old oaks and beeches, in complete contrast to other forests and trees in urban areas, had survived and were still healthy and vibrant. Hambi was, therefore, definitely capable of carrying on. And that put paid to an important argument raised by those who were opposed to halting the clearing—that the forest was no longer worth protecting.

THE HAMBACH FOREST highlighted a completely different conflict, as well: between economic growth and climate protection. For me, this small island of green had become a litmus test for whether politicians really wanted to do anything to reduce carbon dioxide emissions. Forests vacuum up this greenhouse gas and store it in vast quantities in perpetuity. Hambi, for example, stores a total of approximately 200,000 tons, which would be released both directly and indirectly if the forest were cleared. If the lignite beneath it were brought to the surface, of course, it would release much more carbon dioxide than that, which is why it would be better to leave this fuel in the ground.

A few days after our visit, the Higher Administrative Court in Munich ordered a temporary stop to the clearing. The lawsuit brought by the conservation organization Friends of the Earth Germany had succeeded.

WHEN I VISITED Hambi for a second time, in November 2018, I happened across another attempt to cut down trees, this time one that was not in the public spotlight. I was walking with a team from *Stern* magazine to see how things were going. Were the trees still looking healthy? And was this really an ancient woodland? Since the protest, some of the media reports had been trading in half-truths. For instance, Hambi was being portrayed as the last old-growth forest in Germany with trees that were up to twelve thousand years old.

Unfortunately, there is no old-growth forest left in Germany, not a single square foot. The most valuable ecosystems we have left are ancient deciduous forests containing trees that are over three hundred years old—and Hambi counts among them. But not all of it is ancient, as our visit showed. Rows of spruce were evidence of former plantations; birches were evidence of former clear-cuts. The latter cannot compete with beeches and oaks, but they grow well in open spaces. Out in the sun, they can grow more than 3 feet (1 meter) a year and live to be about sixty, a relatively short life for a tree. At that age, beeches and oaks catch up and shade them out.

There are birches in the Hambach Forest today, evidence of human interference in the past. However, large parts of the woodland have returned once again to their natural state thanks to RWE's purchase of the property. The company bought Hambi decades ago in order to get rid of it. It didn't make sense to manage a forest that had been condemned to death, and so Hambi was left in peace. The old oaks grew fatter. Some died and fell over. The woodland became richer in dead wood, a special ecosystem for a few thousand species of insects and fungus. You could say that Hambi fell into an enchanted sleep, like the forest that protected Sleeping Beauty.

As we walked along, the guardians of this sleep rappelled partway down from the canopy. The spokesperson was wearing a mask and introduced himself as Gonzo. A mistrustful fellow eco-warrior wanted to know what we were doing here. When they learned we were writing an article for the magazine, the mood changed, albeit slowly. They agreed to give a brief statement if I would tie a couple of boards lying on the ground to a rope. The woman explained that they wanted to haul them up so they could expand the new tree-house settlement they were building. I was happy to do that, of course, and Gonzo then descended until he was about 10 feet (3 meters) off the ground. He didn't want to come any farther because, in the meantime, a group of workers in yellow safety vests had shown up to—to what? They were not allowed to arrest the activists and there wasn't much building material left lying around (the boards were now on their way up to the canopy). Gonzo called down to them that the press was here, reporting on what was happening. The workers retreated to the jeers of the protesters occupying the trees.

After that, we had an uninterrupted interview with the protesters, visited a second settlement, and took another look over the edge of the open-pit mine just outside the wood. Police officers with dogs milled around every once in a while. I was involuntarily reminded of the border that used to stand between East and West Germany. Just as we left the wood to return to the parking lot, a number of all-terrain vehicles rushed past us toward the tree settlements. The police, who had stopped us shortly before and asked us what we were doing, had probably let them know we had left.

GONZO AND HIS fellow fighters had had a respite of barely three hours. But soon, the break would be a much longer one. On February 1, 2019, the coal commission agreed to announce that they were getting out of the business of coal-fired power plants, and that they would all be shut down by 2038. They had a special recommendation: the Hambach Forest should remain. The federal government signaled that it wanted to abide by their decision. Hambi is here to stay.

31

STRENGTHENING OUR BOND WITH NATURE

YOU'VE NOW LEARNED a few things about trees and our relationship with them. If you'd like to deepen your relationship with these "elephants of the plant world," a change in perspective might be helpful. When people look at a tree, they normally compare the tree with their own body. The crown, because it is at the top, corresponds to the head, then the trunk is the body, and the roots, because they help support the tree, correspond to the feet. That view is mirrored in the technical language, as well, where foresters and scientists talk of the foot of the trunk (at the bottom) and the crown (royals wear them on their heads, as well). However, if there are brain-like structures in the roots, if memories are stored there, and if they carry on clear and effective electric communication with their neighbors via their roots, aren't those organs most comparable with our head, perhaps even with our body?

The part that grows solar cells, the trunk with its branches and leaves, is most like—well, not most like our legs. These areas are where food is produced and processed, where the tree sees, and where it breathes. And this upper part of the tree can regenerate, because many trees grow new shoots when the main trunk is cut down. If, however, you cut away only the roots, the above-ground part of the tree will also die. (Setting aside for now that you could never actually do this because the trunk would no longer have anything to support it.) Given all this, it is more accurate to see the tree as standing on its head. What this perspective does above all is help you better understand these giant beings and have empathy for them. This empathy is extremely important if we want to protect nature.

We can all see the effect of laws and regulations just by looking around: the carbon dioxide content in the air continues to rise, the oceans are full of garbage, and forests are shrinking. A quick course reversal of the kind we need today must come from somewhere else. Think of whales or elephants: we protect them because we have empathy for them. Aren't trees the whales and elephants of the plant world?

As I explained at the beginning of this book, what we need right now is hope, not despair. I have shown you some of the many wonderful ways we are all still part of the natural world and the ways in which we are designed to interact with it. I encourage you to experience this for yourself. Make a plan to go outside and immerse yourself in the sights and sounds of nature.

If there is a forest near you, make that your destination. If you live in a city, find a park or even just a tree-lined street where you can take a walk. On this expedition into the natural world, try to have no fixed destination and no pressing engagements you need to rush back to. Simply take the time you need to engage as many senses as you can. Stand and feel the air on your skin. Is there a soft warm breeze blowing over your arms and legs, or are you all bundled up with just the prick of frosty air on your cheeks? What can you smell? The gentle, earthy aromas of old leaves gently decomposing on the ground or the tangy, brisk scents of new growth? What can you hear? The scratching of squirrels scuttling up trunks or the rustle of leaves as birds turn them over to find insects underneath? What do you see if you look up close? Maybe a spider picking its way across the ridges on a trunk? Or if you look far away? Maybe the pattern of light as sun shines through leaves swaying in the wind? Is your sixth sense at work? Can you sense there is more life out there just hidden from view? Can you sense a change coming in the weather as the breeze freshens or the sun heats up the ground? What about your peripheral vision? Are there other people enjoying the park with you or are you completely on your own? Shut your eyes and feel that this is a place where you belong.

Take a moment to just sit—on a stump or a log or a carpet of leaves. Does that bring you even closer to feeling part of the forest? Run your fingers through the crispness of leaves or over the softness of moss. Be like the children on my hike and don't worry about getting your hands dirty. Forest soil will easily wash off. What do you know about the trees and plants around you? Do you know their names? Do you know if they are safe to eat and, if they are, how they taste? What more would you like to learn about their lives, both what you know you will find in guide books and what you hope scientists will explore in the future so we can really get to know the amazing creatures that are trees in all their biological complexity? We share a world and if they thrive, so do we. We are in this world together, as we always have been, and we are in a position to make sure it stays that way for a long time to come, because this is the world we were made for.

It is by no means too late to protect nature. We are much too tightly bound to it. With the protests in Hambi and in Białowieża, with the school strikes to draw attention to climate change, and with citizens' petitions against honeybee die-off, people have sown the seeds of hope across generations so that now a complete change in direction is being ushered in. A change that is taking place not in our minds but in our hearts.

ACKNOWLEDGMENTS

How do you find all the information you write about in your books? People often ask me this question, and the answer is simple: I'm curious. I come across fascinating phenomena all over the place: in snippets of information in daily newspapers, in conversations with colleagues or scientists, in books, or on my travels. I gather them all together and do further research. I track down the studies behind them, evaluate them, and put all the pieces together to make sense of the puzzle. I combine them with my own observations and sometimes arrive at new conclusions so startling that more than once I've leapt up from my desk and run through the house shouting "I've just got to tell you this. It's just crazy what trees can do!" If you expand the topic from trees to include people and nature, you can imagine how often I absolutely have to tell my wife and children something right away.

And so, from the bottom of my heart, I thank Miriam, Carina, and Tobias for always being there to listen to what I have to say. You also gave me confidence when, although the writing process was well advanced, the manuscript was, as usual, still in a state of complete chaos.

This chaos arose not from lack of planning—there was plenty of that—but from the fact that in the course of my research often not just one but many doors would open. They led to more exciting information, which meant I had to add new chapters and throw out some I had already written. And so the manuscript changed. It grew in some directions and shrank in others, which meant some reorganization was required.

Finally, in February, there was light at the end of the tunnel. I could smooth out the last wrinkles and present the text once more to my wife.

Unlike most readers who are also relatives, and therefore tend to be not critical enough, my wife, Miriam, is very reliable when it comes to giving feedback. She reads through the text even on days when things aren't going so well. She points out the offending passages to me and notes where my storytelling voice begins to waver. On the flip side, I treasure her praise when she's enthusiastic about what I've written. Then I know that I'm headed in the right direction.

Heike Plauert and her team at Ludwig, my German publisher, helped later in the process when I needed to find the right balance between inspiring wonder and relaying information.

In March I had done the work I needed to do on the manuscript, but it was not yet ready to go to print. This was when Angelika Lieke applied herself to my text. It is unbelievable how quickly and unerringly she pounced upon repeated words and gaps in my explanations.

While this was going on, preparations were being made for sales and distribution. The book must physically be there in the bookstores on the day it is published, after all. The printer worked flat out, Beatrice Braken-Gülke prepared television appearances and media interviews, and finally this book saw the light of day. The whole process took two years, and now I'm excited to know what you think about it!

For the English edition, Jane Billinghurst not only did a wonderful job of translating the text but also cleared up a few ambiguous sentences. Finally, Greystone, which has published many of many previous books in English, also published this book. I am very happy about that because it means that the Kwiakah can read what I have written about the days we spent together.

AND WHAT HAPPENED with the other books I wrote? One of the projects I finance with the proceeds is my Forest Academy. It is located in Wershofen on the forested slopes of the Eifel. We offer seminars and short courses with a focus on nature. We also have a research team and a team that supports environmental initiatives. The circle from the forest to the books and back to the forest has now been completed. That makes me happy.

FINALLY, I WOULD especially like to thank the courageous advanced guard of researchers who hold on to their curiosity in the face of entrenched knowledge, ask questions, and dig to find answers that don't fit into the way we have traditionally seen the world. Without these people, my puzzle would have been incomplete and the deciphering of the secret tie that binds people and nature would not have been possible.

NOTES

1. Dale Purves et al. (eds.), "Cones and Color Vision," in *Neuroscience*, 2nd ed. (Sunderland, MA: Sinauer Associates, 2001), www.ncbi.nlm.nih.gov/books/NBK11059.
2. Kim Valenta et al., "The Evolution of Fruit Colour: Phylogeny, Abiotic Factors and the Role of Mutualists," *Scientific Reports* 8, no. 14302 (2018), https://doi.org/10.1038/s41598-018-32604-x.
3. Fiona MacDonald, "There's Evidence Humans Didn't Actually See Blue Until Modern Times," *ScienceAlert*, April 7, 2018, www.sciencealert.com/humans-didn-t-see-the-colour-blue-until-modern-times-evidence-science.
4. Jonathan Winawer et al., "Russian Blues Reveal Effects of Language on Color Discrimination," *Proceedings of the National Academy of Sciences* 104, no. 19 (May 8, 2007): 7780–85, https://doi.org/10.1073/pnas.0701644104.
5. Ian G. Morgan et al., "Myopia," *The Lancet* 379, no. 9872 (May 5, 2012): 1739–48, https://doi.org/10.1016/S0140-6736(12)60272-4.
6. Laura Fademrecht et al., "Action Recognition Is Viewpoint-Dependent in the Visual Periphery," *Vision Research* 135 (June 2017): 10–15, https://dx.doi.org/10.1016/j.visres.2017.01.011.
7. For example, here: https://leswauz.com/2018/06/13/das-faszinierende-hundegehoer-wie-gut-hoert-ein-hund-wirklich/.
8. Sarah Gio, "Wiggle Your Ears? Raise One Eyebrow? The Rare Body Tricks Only Some of Us Can Do," *Glamour* (March 4, 2009), www.glamour.com/story/wiggle-your-ears-raise-one-eye.

9. Kurtis Gruters et al., "The Eardrums Move When the Eyes Move: A Multisensory Effect on the Mechanics of Hearing," *Proceedings of the National Academy of Sciences* 115, no. 6 (February 2018): E1309–18, https://doi.org/10.1073/pnas.1717948115.

10. Martina Stricker, *Mantrailing* (Stuttgart: Franckh Kosmos, 2017), 32.

11. Rolf Froböse, *Wenn Frösche vom Himmel fallen* (When Frogs Fall From the Sky) (Wienheim: Wiley-VCH, 2009).

12. Matthias Laska, "Human and Animal Olfactory Capabilities Compared," in *Springer Handbook of Odor*, ed. Andrea Buettner (Cham, Switzerland: Springer, 2017), https://doi.org/10.1007/978-3-319-26932-0_32.

13. www.augsburger-allgemeine.de/themenwelten/leben-freizeit/Partnersuche-Wie-die-Nase-die-Liebe-bestimmt-id6119146.html.

14. Thomas Braun et al., "Enterochromaffin Cells of the Human Gut: Sensors for Spices and Odorants," *Gastroenterology* 132, no. 5 (2007): 1890–1901, https://doi.org/10.1053/j.gastro.2007.02.036.

15. www.br.de/radio/bayern2/sendungen/iq-wissenschaft-und-forschung/mensch/riechstoerungen-diagnose-therapie100.html.

16. Sonja Steiner-Welz, *Die wichtigsten Körperfunktionen der Menschen* (The Most Important Human Bodily Processes) (Mannheim: Vermittler, 2005), 249.

17. www.tagesspiegel.de/wissen/biologie-auf-den-geschmack-gekommen/1503218.html.

18. Diane Rafizadeh, "The Flow of Flavor: How Exhaling While Eating Affects Smell and Taste," *Yale Scientific* 19 (April 5, 2016): 13, www.yalescientific.org/2016/04/the-flow-of-flavor-how-exhaling-while-eating-affects-smell-and-taste/.

19. Anne C. Gerspach et al., "The Role of the Gut Sweet Taste Receptor in Regulating GLP-1, PYY, and CCK Release in Humans," *American Journal of Physiology-Endocrinology and Metabolism* 301, no. 2 (August 1, 2011): E317–25, https://doi.org/10.1152/ajpendo.00077.2011.

20. "Gut für Gaumen und Verdauung: Forscher entschlüsseln Geheimnis der Gewürze," (Good for Gums and Gut: Scientists Decode the Secrets of Herbs), press release from the Ludwig Maximilian University of Munich, July 11, 2007.

21. Federal Institute for Risk Assessment, "Chemical Food Safety," A/2011 (January 26, 2011), www.bfr.bund.de/en/presseinformation/2011/A/chemical_food_safety-59770.html.

22. Martin Grunwald et al., "Human Haptic Perception Is Interrupted by Explorative Stops of Milliseconds," *Frontiers in Psychology* (April 9, 2014), https://doi.org/10.3389/fpsyg.2014.00292.

23. www.spektrum.de/news/ohne-tastsinn-gibt-es-kein-leben/1302125.

24. www.spektrum.de/news/ohne-tastsinn-gibt-es-kein-leben/1302125.

25. Martin Grunwald et al., "EEG Changes Caused by Spontaneous Facial Self-Touch May Represent Emotion Regulating Processes and Working Memory Maintenance," *Brain Research* 1557 (April 4, 2014): 111–26, https://doi.org/10.1016/j.brainres.2014.02.002.

26. https://rp-online.de/panorama/wissen/der-sechste-sinn-der-tiere_iid-9317101#4.

27. Washington University in St. Louis, "Brain Region Learns to Anticipate Risk, Provides Early Warnings, Suggests New Study," *Science Daily* (February 27, 2005), www.sciencedaily.com/releases/2005/02/050223135039.htm.

28. Karen Kaplan, "Shark Attacks on the Rise? In California, the Risk Has Plunged," *Los Angeles Times*, July 10, 2015, www.latimes.com/science/sciencenow/la-sci-sn-shark-attacks-not-more-frequent-20150708-story.html.

29. Claire G. Williams, "Long-Distance Pine Pollen Still Germinates After Meso-Scale Dispersal," *American Journal of Botany* 97, no. 5 (2010): 846–55, https://doi.org/10.3732/ajb.0900255.

30. www.nabu.de/tiere-und-pflanzen/voegel/vogelkunde/gut-zu-wissen/12017/html.

31. Jörg Fromm and Silke Lautner, "Electrical Signals and Their Physiological Significance in Plants," *Plant, Cell & Environment* 30, no. 3 (March 2007): 249–57, https://doi.org/10.1111/j.1365-3040.2006.01614.x.

32. Ken Yokawa et al., "Anaesthetics Stop Diverse Plant Organ Movements, Affect Endocytic Vesicle Recycling and ROS Homeostasis, and Block Action Potentials in Venus Flytraps," *Annals of Botany* 122, no. 5 (October 5, 2018): 747–56, https://doi.org/10.1093/aob/mcx155.

33. JoAnna Klein, "Sedate a Plant, and It Seems to Lose Consciousness. Is It Conscious?" *New York Times* (February 2, 2018), www.nytimes.com/2018/02/02/science/plants-consciousness-anesthesia.html.

34. www.wissenschaft.de/umwelt-natur/warum-gibt-es-keine-rieseninsekten.

35. Daniel Richter et al., "The Age of the Hominin Fossils from Jebel Irhoud, Morocco, and the Origins of the Middle Stone Age," *Nature* 546 (June 8, 2017): 293–96, https://doi.org/10.1038/nature22335.

36. Chloé Lahondère et al., "What Makes Mosquitoes Attracted to *Platanthera* Orchids?" Abstract for the January 4, 2016, annual meeting of the Society for Integrative and Comparative Biology, http://sicb.org/meetings/2016/schedule/abstractdetails.php?id=349.

37. Mark Bonata et al., "Intentional Firespreading by 'Firehawk' Raptors in Northern Australia," *Journal of Ethnobiology* 34, no. 4 (2017): 700–18, https://doi.org/10.2993/0278-0771-37.4.700.

38. Peter B. Beaumont, "The Edge: More on Fire-Making by About 1.7 Million Years Ago at Wonderwerk Cave in South Africa," *Current Anthropology* 52, no. 4 (August 2011): 585–95, https://doi.org/10.1086/660919.

39. Curtis W. Noonan, Tony J. Ward, and Erin O. Semmens, "Estimating the Number of Vulnerable People in the United States Exposed to Residential Wood Smoke," *Environment Health Perspectives* 123, no. 2 (February 1, 2015): A30–31, https://doi.org/10.1289/ehp.1409136.

40. Troy D. Hubbard et al., "Divergent Ah Receptor Ligand Selectivity During Hominin Evolution," *Molecular Biology and Evolution* 33, no. 10 (October 2016): 2648–58, https://doi.org/10.1093/molbev/msw143.

41. Erica L. Morley and Daniel Robert, "Electric Fields Elicit Ballooning in Spiders," *Current Biology* 28, no. 14 (July 5, 2018): 2324–30, https://doi.org/10.1016/j.cub.2018.05.057.

42. www.wissenschaft.de/umwelt-natur/spannung-liegt-in-der-luft/.

43. Dominic Clarke et al., "Detection and Learning of Floral Electric Fields by Bumblebees," *Science* 340, no. 6128 (April 5, 2013): 66–69, https://doi.org/10.1126/science.1230883.

44. Uwe Greggers et al., "Reception and Learning of Electric Fields in Bees," *Proceedings of the Royal Society B* 280, no. 1759 (May 22, 2013): 20130528, https://doi.org/10.1098/rspb.2013.0528.

45. Michael Kagelidis, "When Does Your Electromagnetic Exposure Exceed the Recommended Safety Limits?" Home Biology, www.home-biology.com/electromagnetic-field-radiation-meters/safe-exposure-limits.

46. Ken-ichi Nakajima et al., "KCNJ15/Kir4.2 Couples With Polyamines to Sense Weak Extracellular Electric Fields in Galvanotaxis," *Nature Communications* 6, no. 8532 (2015), https://doi.org/10.1038/ncomms9532.

47. Federal Office for Radiation Protection, "Recommendations From the BfS for Making Telephone Calls on Mobile Phones," www.bfs.de/EN/topics/emf/mobile-communication/protection/precaution/mobile-phones.html.

48. Peter Schopfer and Axel Brennicke, *Pflanzenphysiologie* (Plant Physiology), 7th ed. (Berlin: Springer, 2016): 585.

49. Elamir Wassim Chehab et al., "*Arabidopsis* Touch-Induced Morphogenesis Is Jasmonate Mediated and Protects Against Pests," *Current Biology* 22, no. 8 (April 24, 2012): 701–6, https://doi.org/10.1016/j.cub.2012.02.061.

50. Eetu Puttonen et al., "Quantification of Overnight Movement of Birch (*Betula pendula*) Branches and Foliage With Short Interval Terrestrial Laser Scanning," *Frontiers in Plant Science* (February 29, 2016), https://doi.org/10.3389/fpls.2016.00222.

51. Andy Coghlan, "Trees May Have a 'Heartbeat' That Is So Slow We Never Noticed It," *New Scientist* (April 20, 2018), www.newscientist.com/article/2167003-trees-may-have-a-heartbeat-that-is-so-slow-we-never-noticed-it/.

52. Ana Rodrigo-Moreno et al., "Root Phonotropism: Early Signaling Events Following Sound Perception in *Arabidopsis* Root," *Plant Science* 264 (November 2017): 9–15, https://doi.org/10.1016/j.plantsci.2017.08.001.

53. Monica Gagliano et al., "Tuned In: Plant Roots Use Sound to Locate Water," *Oecologia* 184, no. 1 (May 2017): 151–60, https://doi.org/10.1007/s00442-017-3862-z.

54. www.planet-wissen.de/natur/pflanzen/sinne_der_pflanzen/pwiewissensfrage528.html.
55. Heidi M. Appel and Reginald B. Cocroft, "Plants Respond to Leaf Vibrations Caused by Insect Herbivore Chewing," *Oecologia* 175, no. 4 (August 2014): 1257–66, https://doi.org/10.1007/s00442-014-2995-6.
56. Paul F. Hendrix et al., "Pandora's Box Contained Bait: The Global Problem of Introduced Earthworms," *Annual Review of Ecology, Evolution, and Systematics* 39 (December 1, 2008): 593–613, http://doi.org/10.1146/annurev.ecolsys.39.110707.173426.
57. Gilbert John and Steffan Messenger, "Ash Dieback: Deadly Tree Fungus Spreading 'More Quickly,'" BBC Wales News (March 8, 2019), www.bbc.com/news/uk-wales-47483197.
58. CAB International, "*Hymenoscyphus fraxineus* (Ash Dieback) Datasheet," www.cabi.org/isc/datasheet/108083.
59. Kelly S. Ramirez et al., "Biogeographic Patterns in Below-Ground Diversity in New York City's Central Park Are Similar to Those Observed Globally," *Proceedings of the Royal Society B* 281, no. 1795 (November 22, 2014): 20141988, https://doi.org/10.1098/rspb.2014.1988.
60. Federal Agency for Nature Conservation, "Wild Alien Animal, Plant and Fungus Species in Germany," www.bfn.de/en/service/facts-and-figures/the-state-of-nature/animals-plants-and-fungi/wild-alien-animal-plant-and-fungus-species-in-germany.html.
61. "The 'Maggio' in Accettura," Basilicata Tourist Board, www.basilicataturistica.it/en/turismi/the-maggio-in-accettura/.
62. Andreas Schneider, *Zypern, DuMont-Reiseführer* (Cyprus, DuMont Travel Guide) (Cologne: DuMont Reise, 2016): 155.
63. "Munlochy Clootie Well," *Explore Inverness*, www.explore-inverness.com/what-to-do/outdoors/munlochy-clootie-well/.
64. www.optik-akademie.com/deu/info-portal/augenoptik/das-auge/die-hornhaut.html.
65. www.baer-linguistick.de/beitrage/jdw/treue.htm.
66. George Monbiot, "Forget 'the Environment': We Need New Words to Convey Life's Wonders," *The Guardian* (August 9, 2017),

https://www.theguardian.com/commentisfree/2017/aug/09/forget-the-environment-new-words-lifes-wonders-language.

67. Katrin Neubauer, "Warum Waldspaziergänge so gesund sind" (Why Forest Walks Are So Healthy), *Spiegel Online* (February 10, 2014), www.spiegel.de/gesundheit/psychologie/waldspaziergaenge-warum-sie-fuer-koerper-und-geist-gesund-sind-a-952492.html.

68. Albert von Haller, *Lebenswichtig aber unerkannt* (Essential but Overlooked) (Langenburg: Boden und Gesundheit, 1980).

69. Christoph Richter, "Phytonzidforschung—ein Beitrag zur Ressourcenfrage" (Phytoncide Research: A Contribution to Our Body of Knowledge), *Hercynia N.F.* 24, no. 1 (1987): 95–106.

70. Janine Fröhlich-Nowoisky et al., "High Diversity of Fungi in Air Particulate Matter," *Proceedings of the National Academy of Sciences* 106, no. 31 (August 4, 2009): 12814–19, https://doi.org/10.1073/pnas.0811003106.

71. Qing Li et al., "Visiting a Forest, but Not a City, Increases Human Natural Killer Activity and Expression of Anti-Cancer Proteins," *International Journal of Immunopathology and Pharmacology* 21, no. 1 (2008): 117–27, https://doi.org/10.1177/039463200802100113.

72. Jee-Yon Lee and Duk-Chul Lee, "Cardiac and Pulmonary Benefits of Forest Walking Versus City Walking in Elderly Women: A Randomised, Controlled, Open-Label Trial," *European Journal of Integrative Medicine* 6, no. 1 (February 2014): 5–11, https://doi.org/10.1016/j.eujim.2013.10.006.

73. Omid Kardan et al., "Neighborhood Greenspace and Health in a Large Urban Center," Scientific *Reports* 5 (2015): 11610, https://doi.org/10.1038/srep11610.

74. Dr. Qing Li, *Shinrin-Yoku* (London: Random House, 2018).

75. https://ihrs.ibe.med.uni-muenchen.de/klimatologie/waldtherapie1.html.

76. Michael A. Huffman, "Animal Self-Medication and Ethno-Medicine: Exploration and Exploitation of the Medicinal Properties of Plants," *Proceedings of the Nutrition Society* 62, no. 2 (May 2003): 371–81, https://doi.org/10.1079/pns2003257.

77. www.spiegel.de/wirtschaft/service/giftpflanze-im-rucola-gestruepp-des-grauens-a-643634.html.

78. Monserrat Suárez-Rodríguez, Isabel López-Rull, and Constantino Macías Garcia, "Incorporation of Cigarette Butts Into Nests Reduces Ectoparasite Load in Urban Birds: New Ingredients for an Old Recipe?" *Biology Letters* 9, no. 1 (February 23, 2013): 20120931, https://doi.org/10.1098/rsbl.2012.0931.
79. https://baumzeitung.de/fileadmin/user_upload/Rinn_Restwand.pdf.
80. Toby Wilkinson, *The Rise and Fall of Ancient Egypt* (New York: Random House, 2010).
81. www.tagesanzeiger.ch/leben/gesellschaft/ist-der-baumder-bessere-mensch/story/29727825.
82. Federal Environment Agency, "Umweltbewusstsein in Deutschland 2016, Ergebnisse einer repräpresentativen Bevölkerungsumfrage" (Environmental Awareness in Germany 2016, Results of a Representative Citizen Survey), April 2016.
83. World Health Organization, "Night Noise Guidelines for Europe," 2009, http://www.euro.who.int/__data/assets/pdf_file/0017/43316/E92845.pdf?ua=1.
84. European Environment Agency, "Air Quality in Europe—2018 Report" (Luxembourg: Publications Office of the European Union, 2018), https://doi.org/10.2800/777411.
85. www.umwelbundesamt.de/themen/wirstschaft-konsum/industriebranchen/feuerungsanlagen/kleine-mittlere-feuerungsanlagen#textpart-1.
86. Fiona Harvey, "Muck-Spreading Could Be Banned to Reduce Air Pollution," *The Guardian* (January 14, 2019), www.theguardian.com/environment/2019/jan/14/muck-spreading-could-be-banned-to-reduce-air-pollution.
87. Frank Harmuth, *Der sächsische Wald im Dienst der Allgemeinheit* (Woodlands in Saxony in Service of the Public) (Pirna: Staatsbetrieb Sachsenforst, October 2003): 33.
88. www.bussgeldkatalog.org/umwelt-baum-faellen.
89. Kim Naudts et al., "Europe's Forest Management Did Not Mitigate Climate Warming," *Science* 351, no. 6273 (February 5, 2016): 597, https://doi.org/10.1126/science.aad7270.

90. https://de.statista.com/statistik/daten/studie/179260/umfrage/die-zehn-groessten-co2-emittenten-weltweit/.

91. Markus Rex from the Alfred Wegener Institute on ice melt in the Arctic, radioWelt (radio program), October 26, 2018, Bavaria 2.

92. Michon Scott, "February 2018 Heatwave Across the Far North," National Oceanic and Atmospheric Administration Climate.gov, March 20, 2018, www.climate.gov/news-features/event-tracker/february-2018-heatwave-across-far-north.

93. Ingmar Nitze, Climate change in the Artic: a conversation with Arndt Reuning, Deutschlandfunk (radio program), July 18, 2018.

94. Simon Evans, "Investigation: Does the UK's Biomass Burning Help Solve Climate Change?" *Carbon Brief*, May 11, 2015, www.carbonbrief.org/investigation-does-the-uks-biomass-burning-help-solve-climate-change.

95. https://plattform-wald-klima.de/2019/02/20/fake-news-oder-klimaloesung-drax-will-englische-biertrinker-zu-klimaschuetzern-machen/.

96. Indigenous Training Inc., "Indigenous Culturally Modified Trees," September 10, 2019, www.ictinc.ca/blog/indigenous-culturally-modified-trees.

INDEX